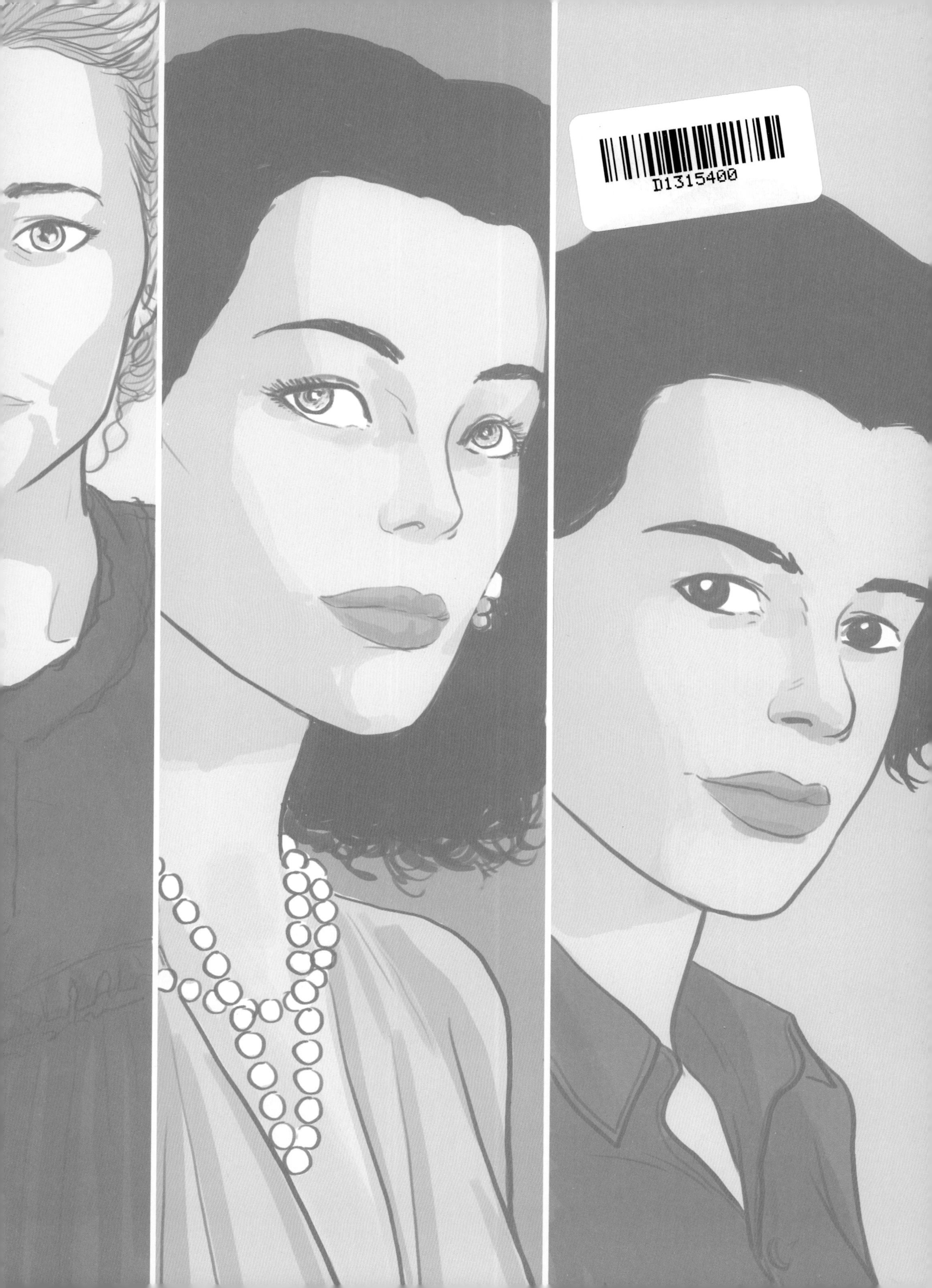

WOMEN DISCOVERERS

TOP WOMEN IN SCIENCE

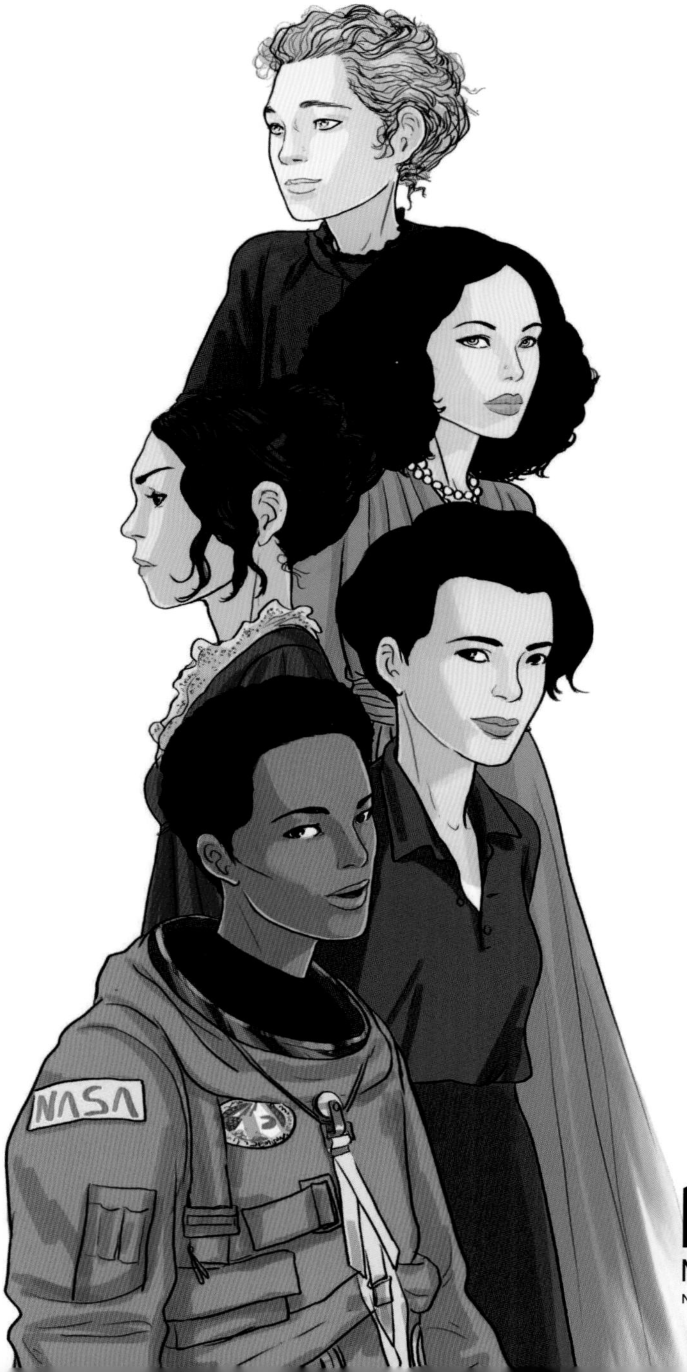
NASA

nbm GRAPHIC NOVELS
Nantier • Beall • Minoustchine
NEW YORK

Thanks to Christelle for having put these pages into pictures so beautifully. This is for girls to feel able to dare all their dreams.
To Noémie.
Marie Moinard

Thanks to Marie to have left me the freedom I needed, essential to reach our goals. And to Jean-Paul Moulin for his patience.
To all women who didn't always get the recognition they deserved for their work.
Christelle Pecout

Thanks to Lucie and Iris Moulin for having been the inspiration behind this project and for their sustained support.
Jean-Paul Moulin

ISBN 9781681122700

Library of Congress Control Number: 2020950798
Translation by Montana Kane
Lettering by Ortho
Printed in China
1st printing March 2021

This book is also available wherever e-books are sold
ISBN 9781681122717

nbm
GRAPHIC NOVELS
Comics Biographies

INTRODUCTION

"The bells never sound the end of discovery." -Colette

While countless women throughout history have made enormous contributions to the fields of science and technology, many of them, unfortunately, are not synonymous with the words "discovery" and "invention" in the minds of the general public.

From Ada Lovelace to Xie Yi, from Marie Curie to Emilie du Châtelet, this book is so fascinating you will want to read it a single sitting! It will take you on a journey around the world on which you will be introduced to some of the Great Ladies of Science and Technology from the last few centuries!

With this book, equally remarkable in its writing and its illustrations, authors Marie Moinard and Christelle Pecout pay a beautiful tribute to these emblematic women who left their mark on the world with wide-reaching scientific discoveries and were often pioneers of cutting-edge technologies currently used by our businesses and economies 4.0.

As you will discover, their work and research was sometimes attributed to their male colleagues or remained in the shadows for years...

Fortunately, mindsets have changed over the years and the time has come to shine the spotlight on these exceptional women and their extraordinary accomplishments, who were always driven by a passion for science and a deep desire to help advance research and technology.

Our twenty-first century society needs to promote awareness of these female role models, so that their accomplishments can inspire more and more young women to follow the calling of science and go on to make discoveries of their own.

To that end, reading this Women Discoverers, Top Women in Science is a great place to start.

Happy reading!

Marie- Sophie Pawlak
President of "Elles Bougent," a French society for the promotion of women in Science

TABLE OF CONTENTS

MARIE CURIE

1867 - 1934

Polish-French physicist and chemist who, along with her husband Pierre Currie, discovered radium. Nobel Prize in Physics 1903 shared with Pierre Curie and Henri Becquerel. Nobel Prize in Chemistry 1911.

PARIS – 1891.

MARIA, MY DEAR SISTER!

BRONIA!

WE DID IT! I NEVER COULD HAVE COME TO PARIS TO STUDY WITHOUT THE MONEY YOU SENT. YOU WORKED SO HARD TO KEEP YOUR PROMISE!

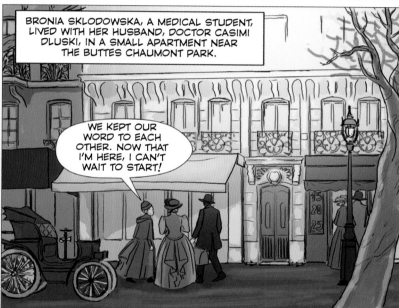

BRONIA SKLODOWSKA, A MEDICAL STUDENT, LIVED WITH HER HUSBAND, DOCTOR CASIMI DLUSKI, IN A SMALL APARTMENT NEAR THE BUTTES CHAUMONT PARK.

WE KEPT OUR WORD TO EACH OTHER. NOW THAT I'M HERE, I CAN'T WAIT TO START!

WELL DONE, MARIA! YOU BROKE ALL THE RULES BY ATTENDING THE SECRET CLASSES FOR WOMEN AT THE FLYING UNIVERSITY OF KRAKOW.

I HAD NO CHOICE, GIVEN THAT THE RUSSIAN EMPIRE BANS YOUNG POLISH WOMEN FROM HIGHER EDUCATION.

WELL, YOU'RE IN PARIS NOW, AND THE SORBONNE AWAITS. LET'S GO GET YOU ENROLLED!

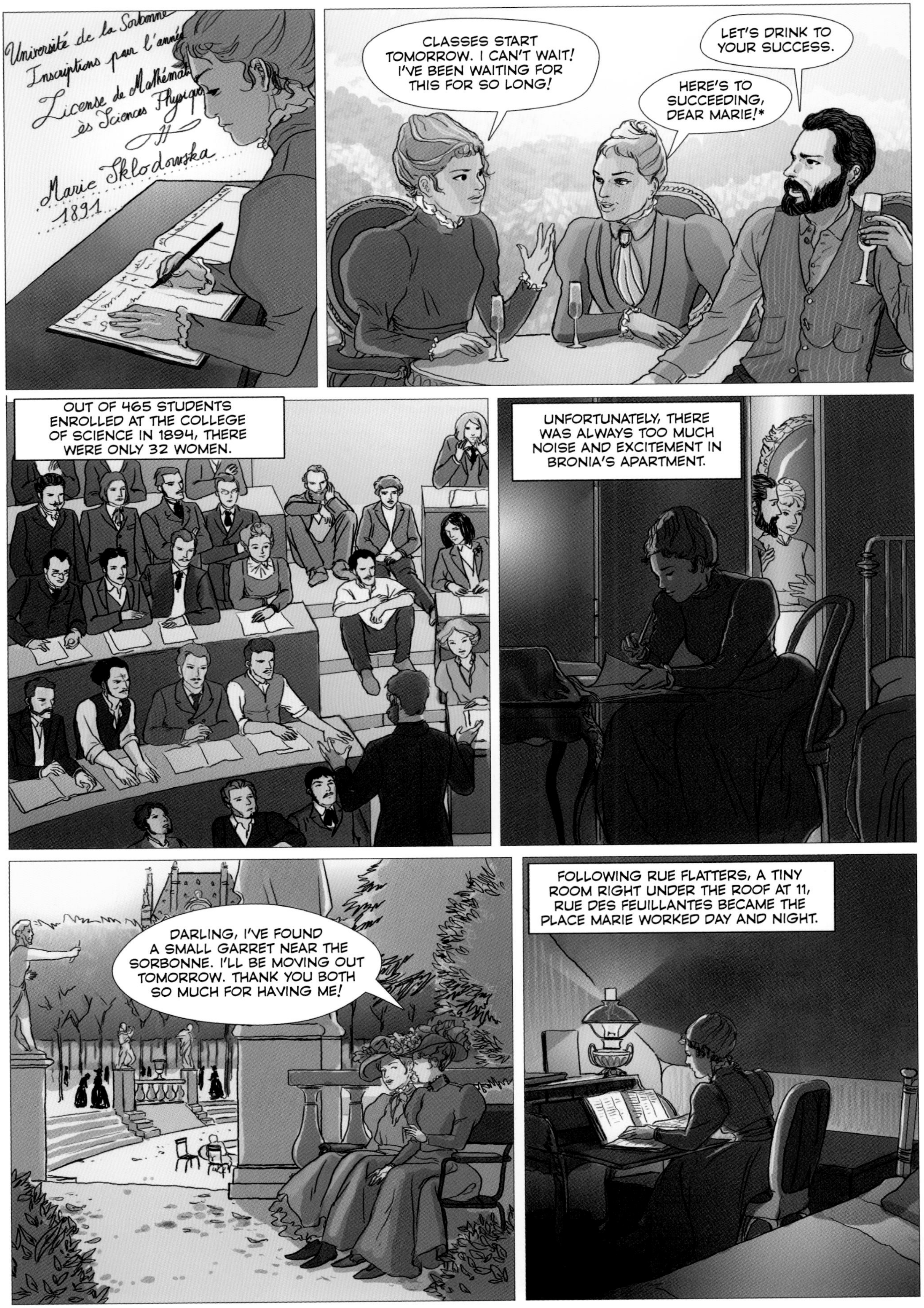
Université de la Sorbonne
Inscriptions pour l'anné
License de Mathémat
ès Sciences Physiqu
Marie Sklodowska
1891
CLASSES START TOMORROW. I CAN'T WAIT! I'VE BEEN WAITING FOR THIS FOR SO LONG!
HERE'S TO SUCCEEDING, DEAR MARIE!*
LET'S DRINK TO YOUR SUCCESS.
OUT OF 465 STUDENTS ENROLLED AT THE COLLEGE OF SCIENCE IN 1894, THERE WERE ONLY 32 WOMEN.
UNFORTUNATELY, THERE WAS ALWAYS TOO MUCH NOISE AND EXCITEMENT IN BRONIA'S APARTMENT.
DARLING, I'VE FOUND A SMALL GARRET NEAR THE SORBONNE. I'LL BE MOVING OUT TOMORROW. THANK YOU BOTH SO MUCH FOR HAVING ME!
FOLLOWING RUE FLATTERS, A TINY ROOM RIGHT UNDER THE ROOF AT 11, RUE DES FEUILLANTES BECAME THE PLACE MARIE WORKED DAY AND NIGHT.

* MARIA BECAME MARIE WHEN SHE ENROLLED AT THE SORBONNE.

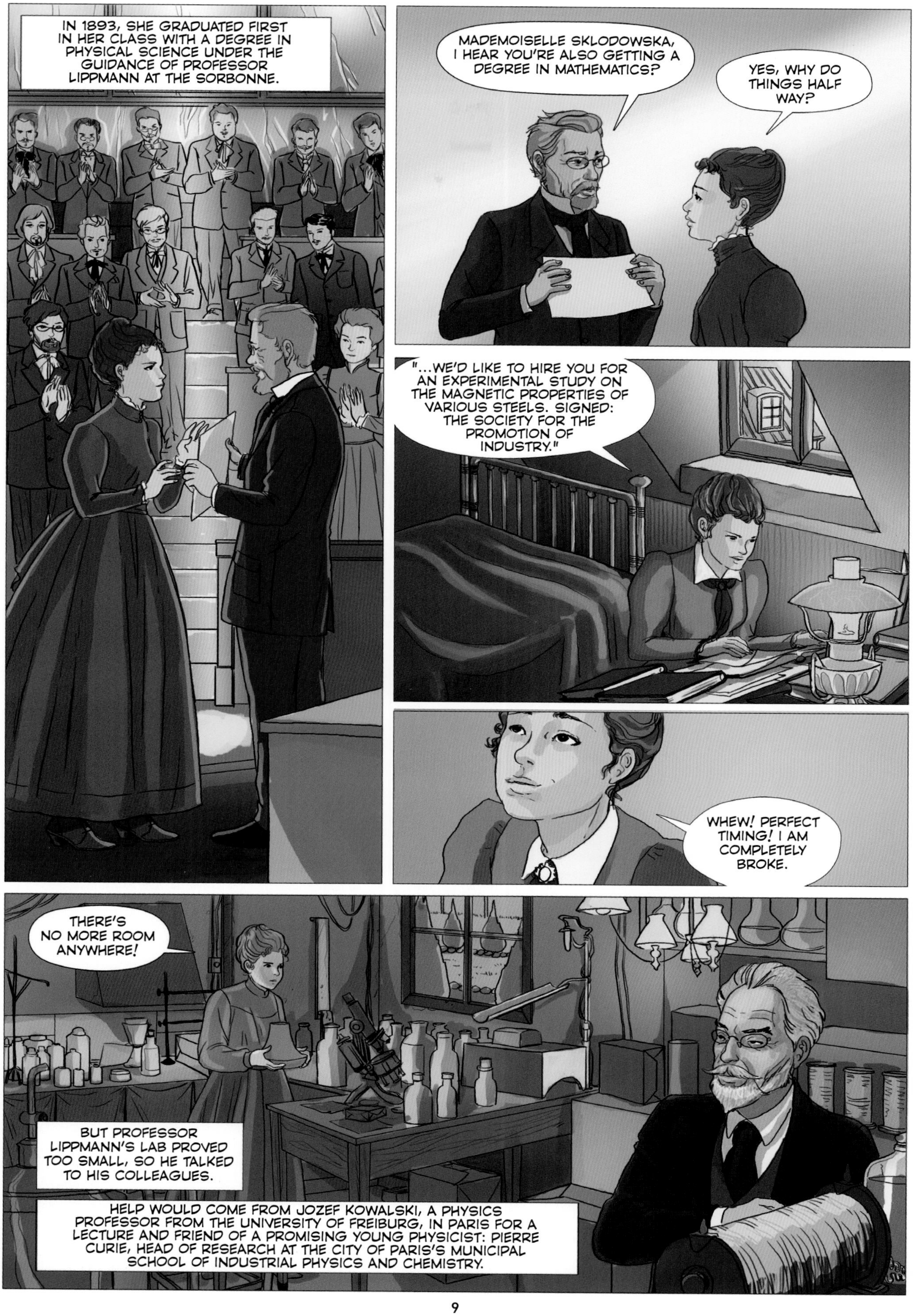
IN 1893, SHE GRADUATED FIRST IN HER CLASS WITH A DEGREE IN PHYSICAL SCIENCE UNDER THE GUIDANCE OF PROFESSOR LIPPMANN AT THE SORBONNE.
MADEMOISELLE SKLODOWSKA, I HEAR YOU'RE ALSO GETTING A DEGREE IN MATHEMATICS?
YES, WHY DO THINGS HALF WAY?
"...WE'D LIKE TO HIRE YOU FOR AN EXPERIMENTAL STUDY ON THE MAGNETIC PROPERTIES OF VARIOUS STEELS. SIGNED: THE SOCIETY FOR THE PROMOTION OF INDUSTRY."
WHEW! PERFECT TIMING! I AM COMPLETELY BROKE.
THERE'S NO MORE ROOM ANYWHERE!
BUT PROFESSOR LIPPMANN'S LAB PROVED TOO SMALL, SO HE TALKED TO HIS COLLEAGUES.
HELP WOULD COME FROM JOZEF KOWALSKI, A PHYSICS PROFESSOR FROM THE UNIVERSITY OF FREIBURG, IN PARIS FOR A LECTURE AND FRIEND OF A PROMISING YOUNG PHYSICIST: PIERRE CURIE, HEAD OF RESEARCH AT THE CITY OF PARIS'S MUNICIPAL SCHOOL OF INDUSTRIAL PHYSICS AND CHEMISTRY.

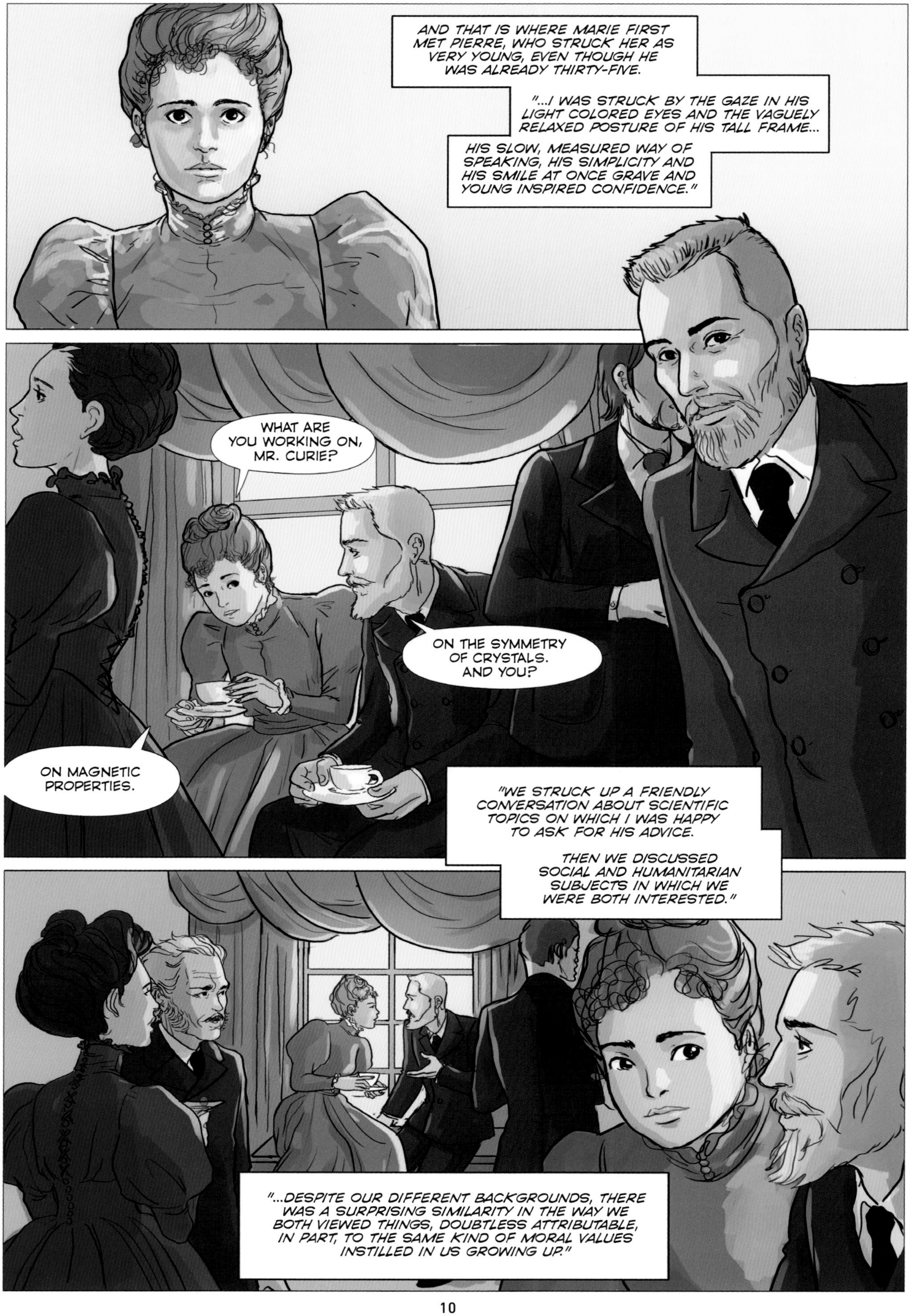
AND THAT IS WHERE MARIE FIRST MET PIERRE, WHO STRUCK HER AS VERY YOUNG, EVEN THOUGH HE WAS ALREADY THIRTY-FIVE.
"...I WAS STRUCK BY THE GAZE IN HIS LIGHT COLORED EYES AND THE VAGUELY RELAXED POSTURE OF HIS TALL FRAME...
HIS SLOW, MEASURED WAY OF SPEAKING, HIS SIMPLICITY AND HIS SMILE AT ONCE GRAVE AND YOUNG INSPIRED CONFIDENCE."
WHAT ARE YOU WORKING ON, MR. CURIE?
ON THE SYMMETRY OF CRYSTALS. AND YOU?
ON MAGNETIC PROPERTIES.
"WE STRUCK UP A FRIENDLY CONVERSATION ABOUT SCIENTIFIC TOPICS ON WHICH I WAS HAPPY TO ASK FOR HIS ADVICE.
THEN WE DISCUSSED SOCIAL AND HUMANITARIAN SUBJECTS IN WHICH WE WERE BOTH INTERESTED."
"...DESPITE OUR DIFFERENT BACKGROUNDS, THERE WAS A SURPRISING SIMILARITY IN THE WAY WE BOTH VIEWED THINGS, DOUBTLESS ATTRIBUTABLE, IN PART, TO THE SAME KIND OF MORAL VALUES INSTILLED IN US GROWING UP."

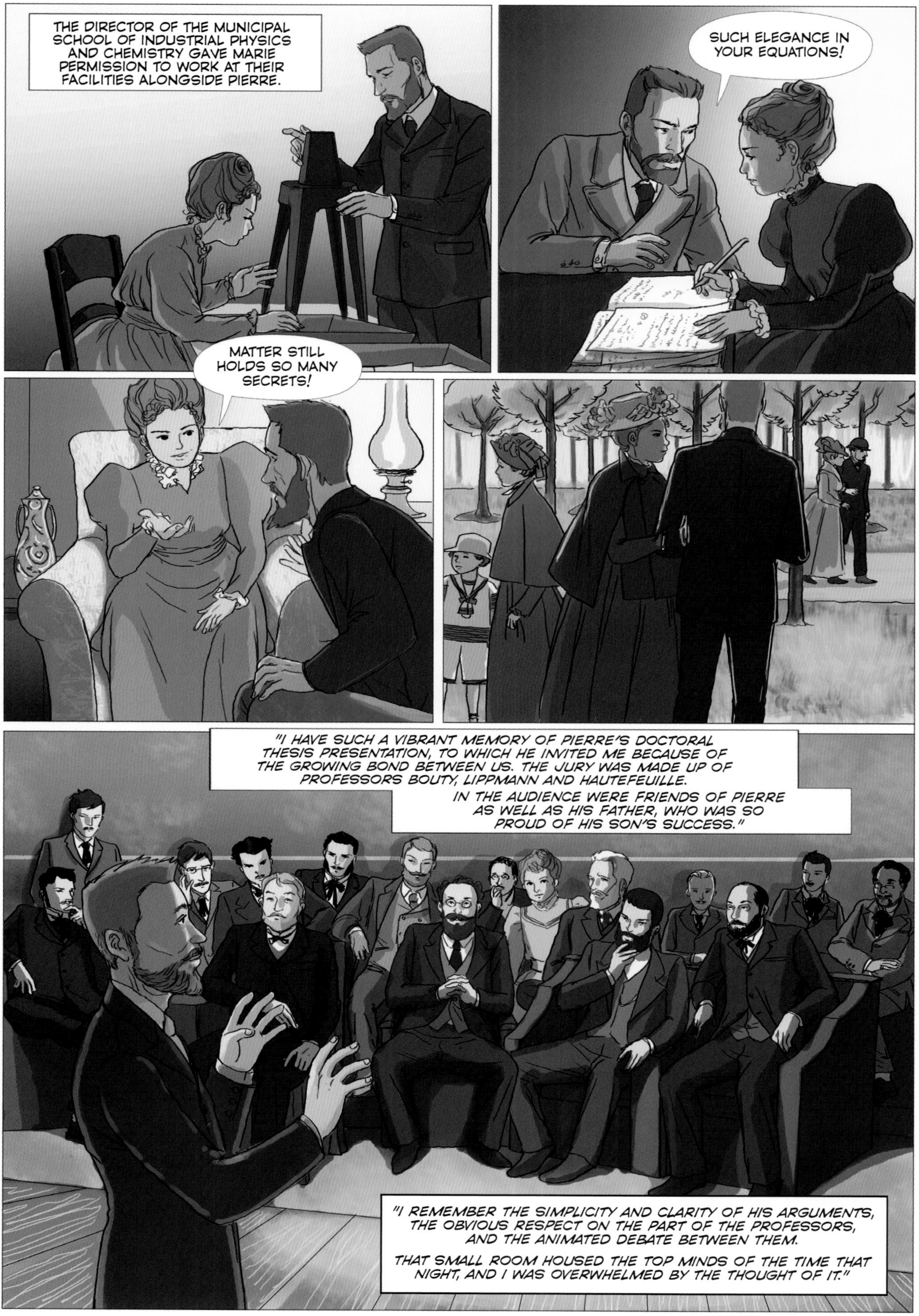

THE DIRECTOR OF THE MUNICIPAL SCHOOL OF INDUSTRIAL PHYSICS AND CHEMISTRY GAVE MARIE PERMISSION TO WORK AT THEIR FACILITIES ALONGSIDE PIERRE.
SUCH ELEGANCE IN YOUR EQUATIONS!
MATTER STILL HOLDS SO MANY SECRETS!
"I HAVE SUCH A VIBRANT MEMORY OF PIERRE'S DOCTORAL THESIS PRESENTATION, TO WHICH HE INVITED ME BECAUSE OF THE GROWING BOND BETWEEN US. THE JURY WAS MADE UP OF PROFESSORS BOUTY, LIPPMANN AND HAUTEFEUILLE.
IN THE AUDIENCE WERE FRIENDS OF PIERRE AS WELL AS HIS FATHER, WHO WAS SO PROUD OF HIS SON'S SUCCESS."
"I REMEMBER THE SIMPLICITY AND CLARITY OF HIS ARGUMENTS, THE OBVIOUS RESPECT ON THE PART OF THE PROFESSORS, AND THE ANIMATED DEBATE BETWEEN THEM.
THAT SMALL ROOM HOUSED THE TOP MINDS OF THE TIME THAT NIGHT, AND I WAS OVERWHELMED BY THE THOUGHT OF IT."

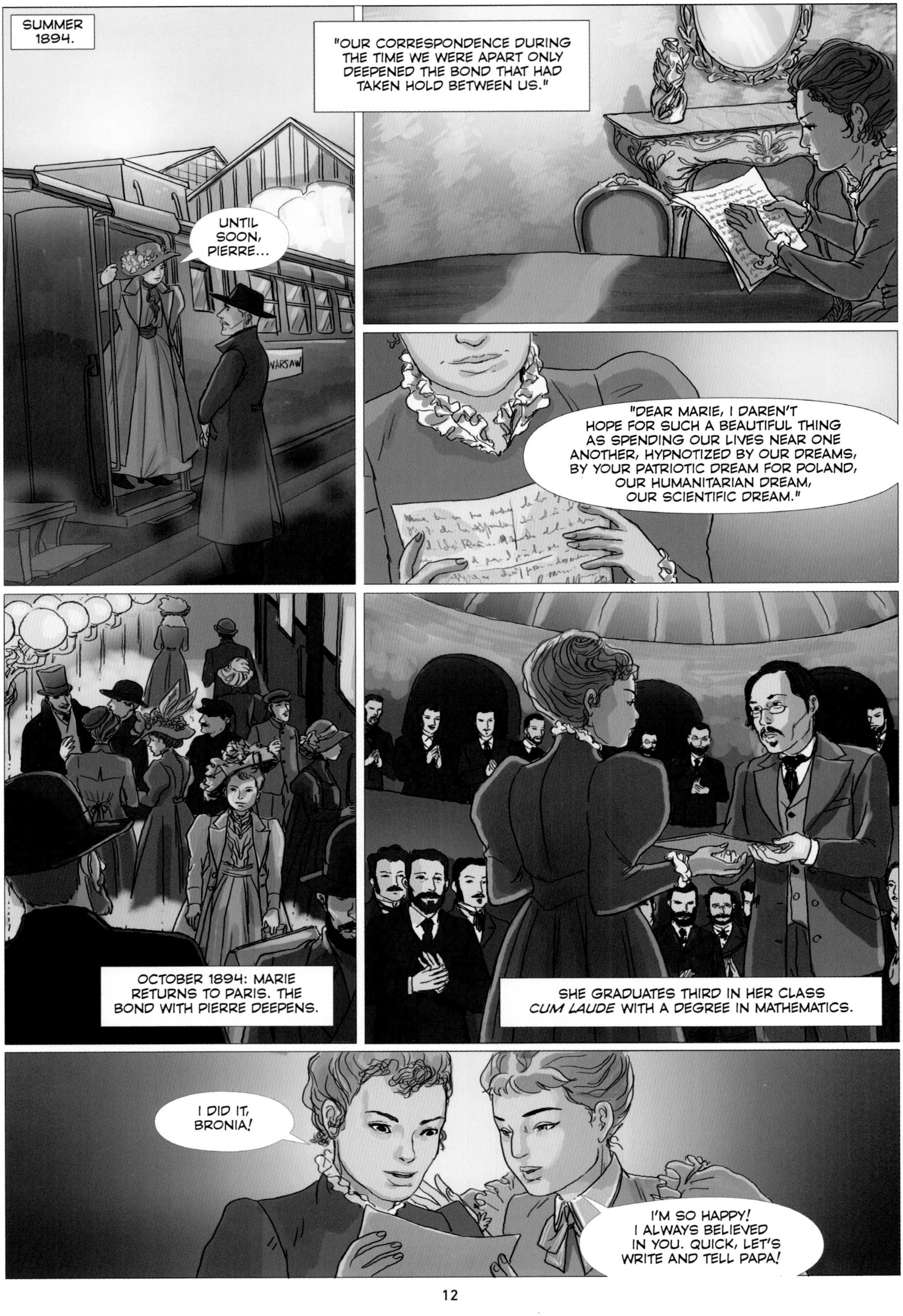
SUMMER 1894.
UNTIL SOON, PIERRE...
WARSAW
"OUR CORRESPONDENCE DURING THE TIME WE WERE APART ONLY DEEPENED THE BOND THAT HAD TAKEN HOLD BETWEEN US."
"DEAR MARIE, I DAREN'T HOPE FOR SUCH A BEAUTIFUL THING AS SPENDING OUR LIVES NEAR ONE ANOTHER, HYPNOTIZED BY OUR DREAMS, BY YOUR PATRIOTIC DREAM FOR POLAND, OUR HUMANITARIAN DREAM, OUR SCIENTIFIC DREAM."
OCTOBER 1894: MARIE RETURNS TO PARIS. THE BOND WITH PIERRE DEEPENS.
SHE GRADUATES THIRD IN HER CLASS *CUM LAUDE* WITH A DEGREE IN MATHEMATICS.
I DID IT, BRONIA!
I'M SO HAPPY! I ALWAYS BELIEVED IN YOU. QUICK, LET'S WRITE AND TELL PAPA!

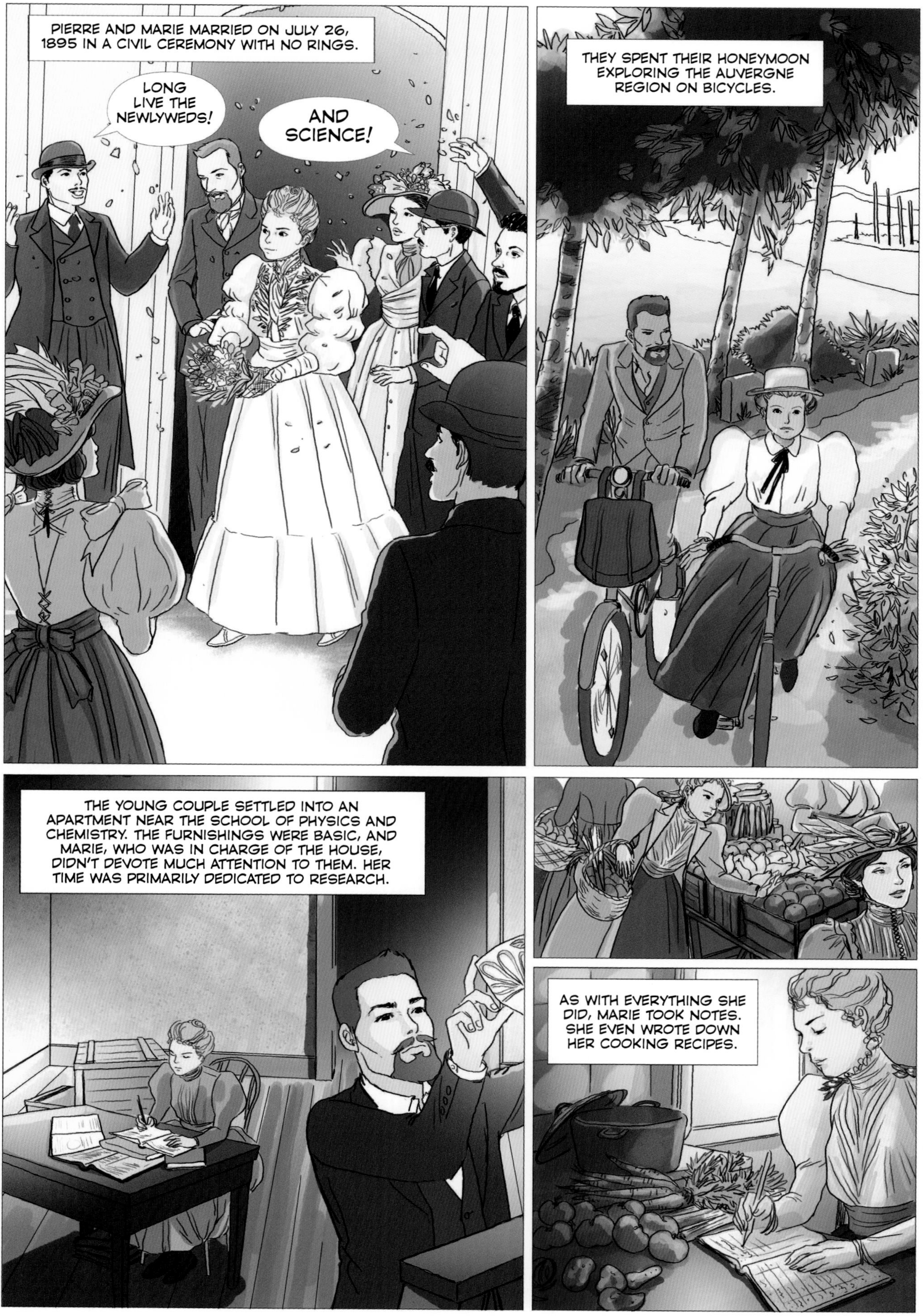
PIERRE AND MARIE MARRIED ON JULY 26, 1895 IN A CIVIL CEREMONY WITH NO RINGS.
LONG LIVE THE NEWLYWEDS!
AND SCIENCE!
THEY SPENT THEIR HONEYMOON EXPLORING THE AUVERGNE REGION ON BICYCLES.
THE YOUNG COUPLE SETTLED INTO AN APARTMENT NEAR THE SCHOOL OF PHYSICS AND CHEMISTRY. THE FURNISHINGS WERE BASIC, AND MARIE, WHO WAS IN CHARGE OF THE HOUSE, DIDN'T DEVOTE MUCH ATTENTION TO THEM. HER TIME WAS PRIMARILY DEDICATED TO RESEARCH.
AS WITH EVERYTHING SHE DID, MARIE TOOK NOTES. SHE EVEN WROTE DOWN HER COOKING RECIPES.

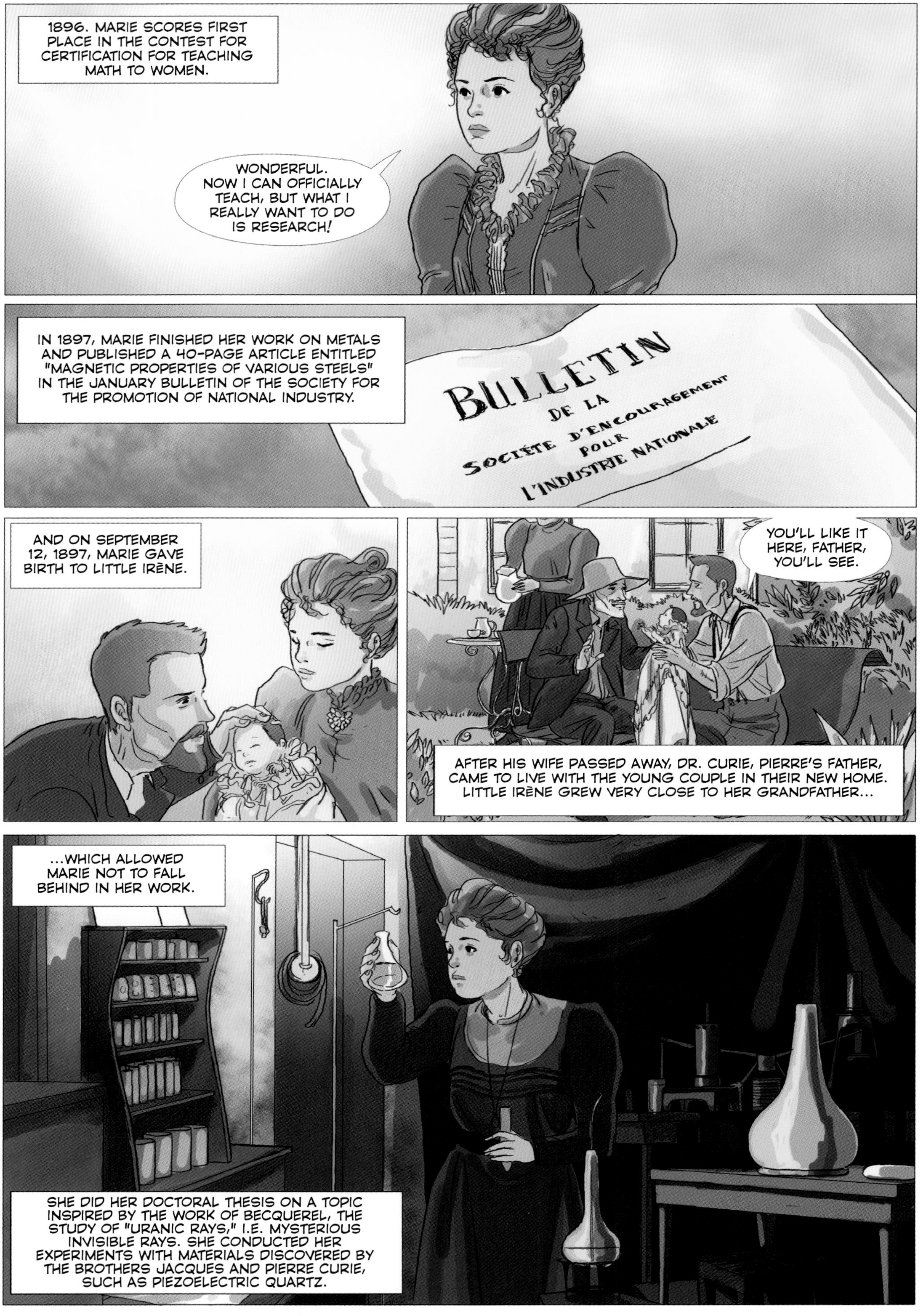
1896. MARIE SCORES FIRST PLACE IN THE CONTEST FOR CERTIFICATION FOR TEACHING MATH TO WOMEN.
WONDERFUL. NOW I CAN OFFICIALLY TEACH, BUT WHAT I REALLY WANT TO DO IS RESEARCH!
IN 1897, MARIE FINISHED HER WORK ON METALS AND PUBLISHED A 40-PAGE ARTICLE ENTITLED "MAGNETIC PROPERTIES OF VARIOUS STEELS" IN THE JANUARY BULLETIN OF THE SOCIETY FOR THE PROMOTION OF NATIONAL INDUSTRY.
BULLETIN DE LA SOCIÉTE D'ENCOURAGEMENT POUR L'INDUSTRIE NATIONALE
AND ON SEPTEMBER 12, 1897, MARIE GAVE BIRTH TO LITTLE IRÈNE.
YOU'LL LIKE IT HERE, FATHER, YOU'LL SEE.
AFTER HIS WIFE PASSED AWAY, DR. CURIE, PIERRE'S FATHER, CAME TO LIVE WITH THE YOUNG COUPLE IN THEIR NEW HOME. LITTLE IRÈNE GREW VERY CLOSE TO HER GRANDFATHER...
...WHICH ALLOWED MARIE NOT TO FALL BEHIND IN HER WORK.
SHE DID HER DOCTORAL THESIS ON A TOPIC INSPIRED BY THE WORK OF BECQUEREL, THE STUDY OF "URANIC RAYS," I.E. MYSTERIOUS INVISIBLE RAYS. SHE CONDUCTED HER EXPERIMENTS WITH MATERIALS DISCOVERED BY THE BROTHERS JACQUES AND PIERRE CURIE, SUCH AS PIEZOELECTRIC QUARTZ.

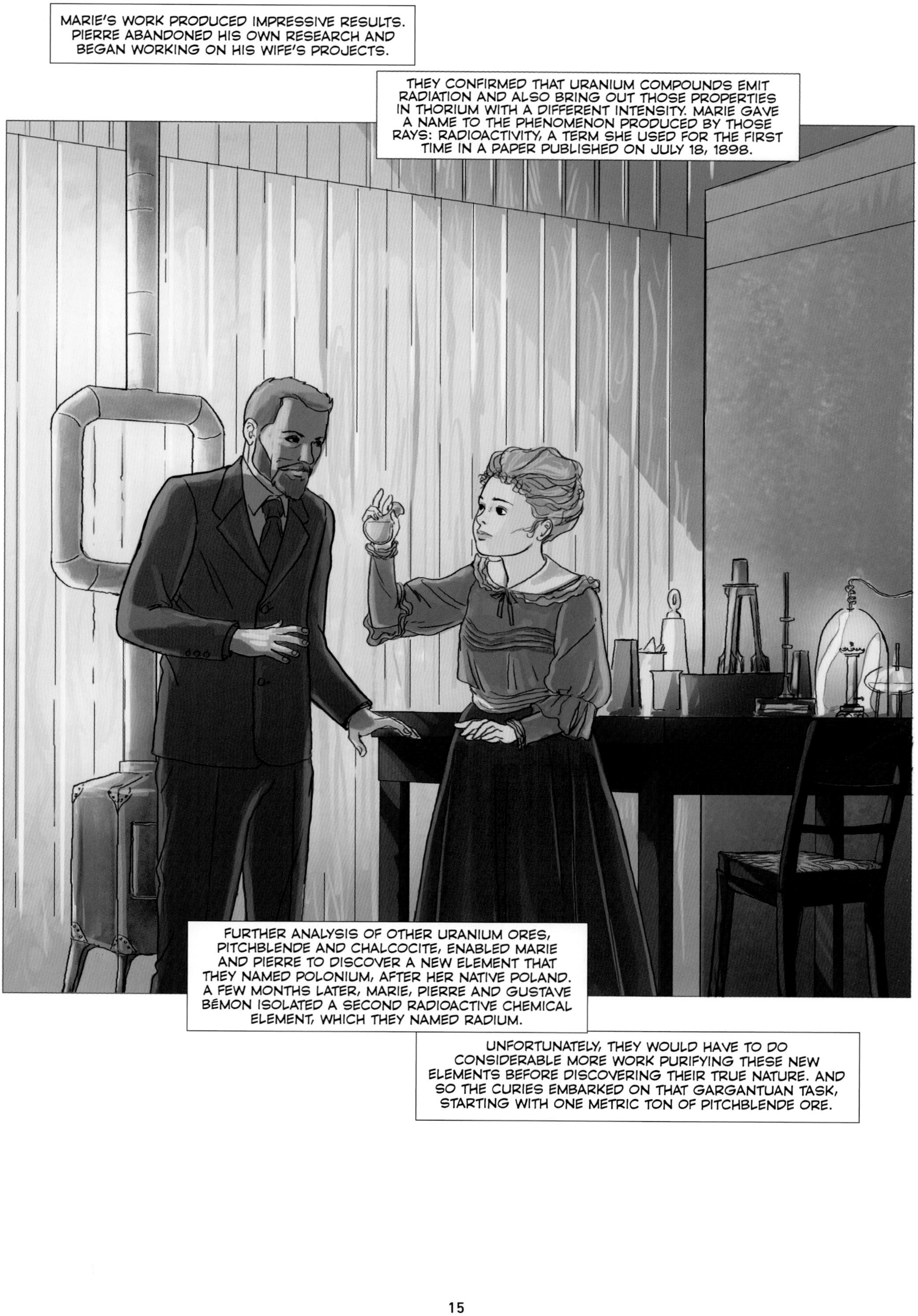
MARIE'S WORK PRODUCED IMPRESSIVE RESULTS. PIERRE ABANDONED HIS OWN RESEARCH AND BEGAN WORKING ON HIS WIFE'S PROJECTS.
THEY CONFIRMED THAT URANIUM COMPOUNDS EMIT RADIATION AND ALSO BRING OUT THOSE PROPERTIES IN THORIUM WITH A DIFFERENT INTENSITY. MARIE GAVE A NAME TO THE PHENOMENON PRODUCED BY THOSE RAYS: RADIOACTIVITY, A TERM SHE USED FOR THE FIRST TIME IN A PAPER PUBLISHED ON JULY 18, 1898.
FURTHER ANALYSIS OF OTHER URANIUM ORES, PITCHBLENDE AND CHALCOCITE, ENABLED MARIE AND PIERRE TO DISCOVER A NEW ELEMENT THAT THEY NAMED POLONIUM, AFTER HER NATIVE POLAND. A FEW MONTHS LATER, MARIE, PIERRE AND GUSTAVE BÉMON ISOLATED A SECOND RADIOACTIVE CHEMICAL ELEMENT, WHICH THEY NAMED RADIUM.
UNFORTUNATELY, THEY WOULD HAVE TO DO CONSIDERABLE MORE WORK PURIFYING THESE NEW ELEMENTS BEFORE DISCOVERING THEIR TRUE NATURE. AND SO THE CURIES EMBARKED ON THAT GARGANTUAN TASK, STARTING WITH ONE METRIC TON OF PITCHBLENDE ORE.

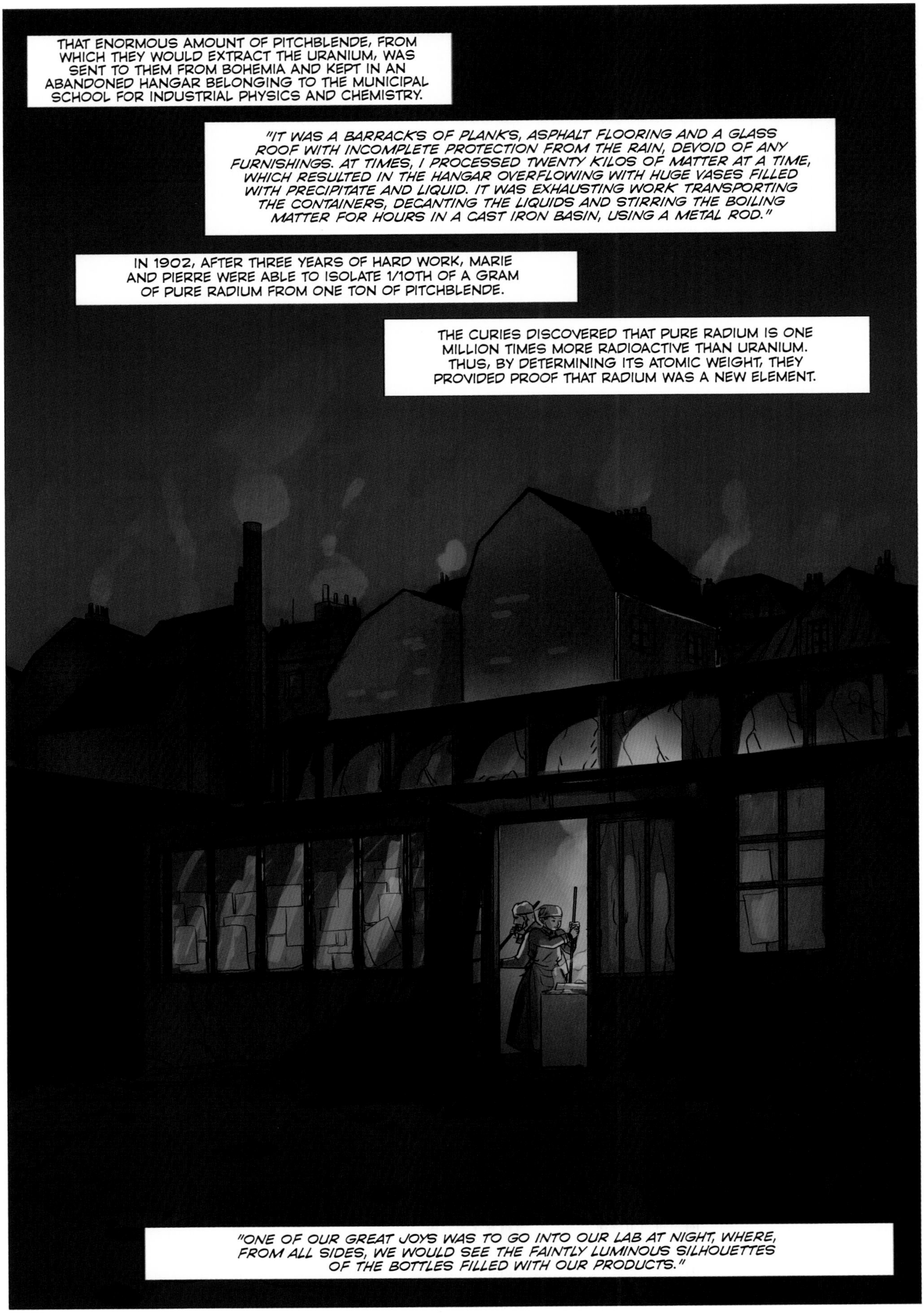
THAT ENORMOUS AMOUNT OF PITCHBLENDE, FROM WHICH THEY WOULD EXTRACT THE URANIUM, WAS SENT TO THEM FROM BOHEMIA AND KEPT IN AN ABANDONED HANGAR BELONGING TO THE MUNICIPAL SCHOOL FOR INDUSTRIAL PHYSICS AND CHEMISTRY.
"IT WAS A BARRACKS OF PLANKS, ASPHALT FLOORING AND A GLASS ROOF WITH INCOMPLETE PROTECTION FROM THE RAIN, DEVOID OF ANY FURNISHINGS. AT TIMES, I PROCESSED TWENTY KILOS OF MATTER AT A TIME, WHICH RESULTED IN THE HANGAR OVERFLOWING WITH HUGE VASES FILLED WITH PRECIPITATE AND LIQUID. IT WAS EXHAUSTING WORK TRANSPORTING THE CONTAINERS, DECANTING THE LIQUIDS AND STIRRING THE BOILING MATTER FOR HOURS IN A CAST IRON BASIN, USING A METAL ROD."
IN 1902, AFTER THREE YEARS OF HARD WORK, MARIE AND PIERRE WERE ABLE TO ISOLATE 1/10TH OF A GRAM OF PURE RADIUM FROM ONE TON OF PITCHBLENDE.
THE CURIES DISCOVERED THAT PURE RADIUM IS ONE MILLION TIMES MORE RADIOACTIVE THAN URANIUM. THUS, BY DETERMINING ITS ATOMIC WEIGHT, THEY PROVIDED PROOF THAT RADIUM WAS A NEW ELEMENT.
"ONE OF OUR GREAT JOYS WAS TO GO INTO OUR LAB AT NIGHT, WHERE, FROM ALL SIDES, WE WOULD SEE THE FAINTLY LUMINOUS SILHOUETTES OF THE BOTTLES FILLED WITH OUR PRODUCTS."

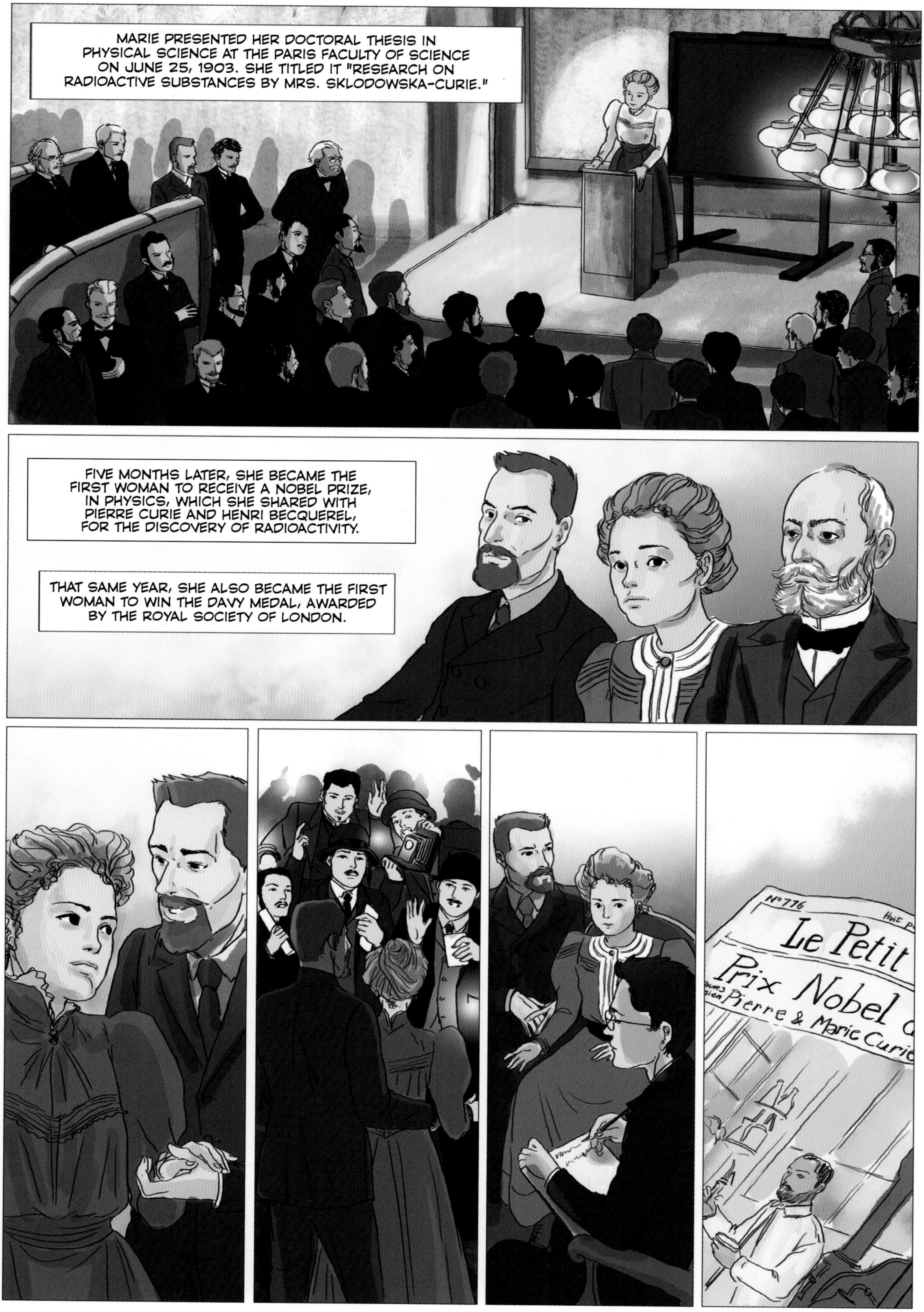
MARIE PRESENTED HER DOCTORAL THESIS IN PHYSICAL SCIENCE AT THE PARIS FACULTY OF SCIENCE ON JUNE 25, 1903. SHE TITLED IT "RESEARCH ON RADIOACTIVE SUBSTANCES BY MRS. SKLODOWSKA-CURIE."
FIVE MONTHS LATER, SHE BECAME THE FIRST WOMAN TO RECEIVE A NOBEL PRIZE, IN PHYSICS, WHICH SHE SHARED WITH PIERRE CURIE AND HENRI BECQUEREL, FOR THE DISCOVERY OF RADIOACTIVITY.
THAT SAME YEAR, SHE ALSO BECAME THE FIRST WOMAN TO WIN THE DAVY MEDAL, AWARDED BY THE ROYAL SOCIETY OF LONDON.
N°776
Le Petit
Prix Nobel
Pierre & Marie Curie

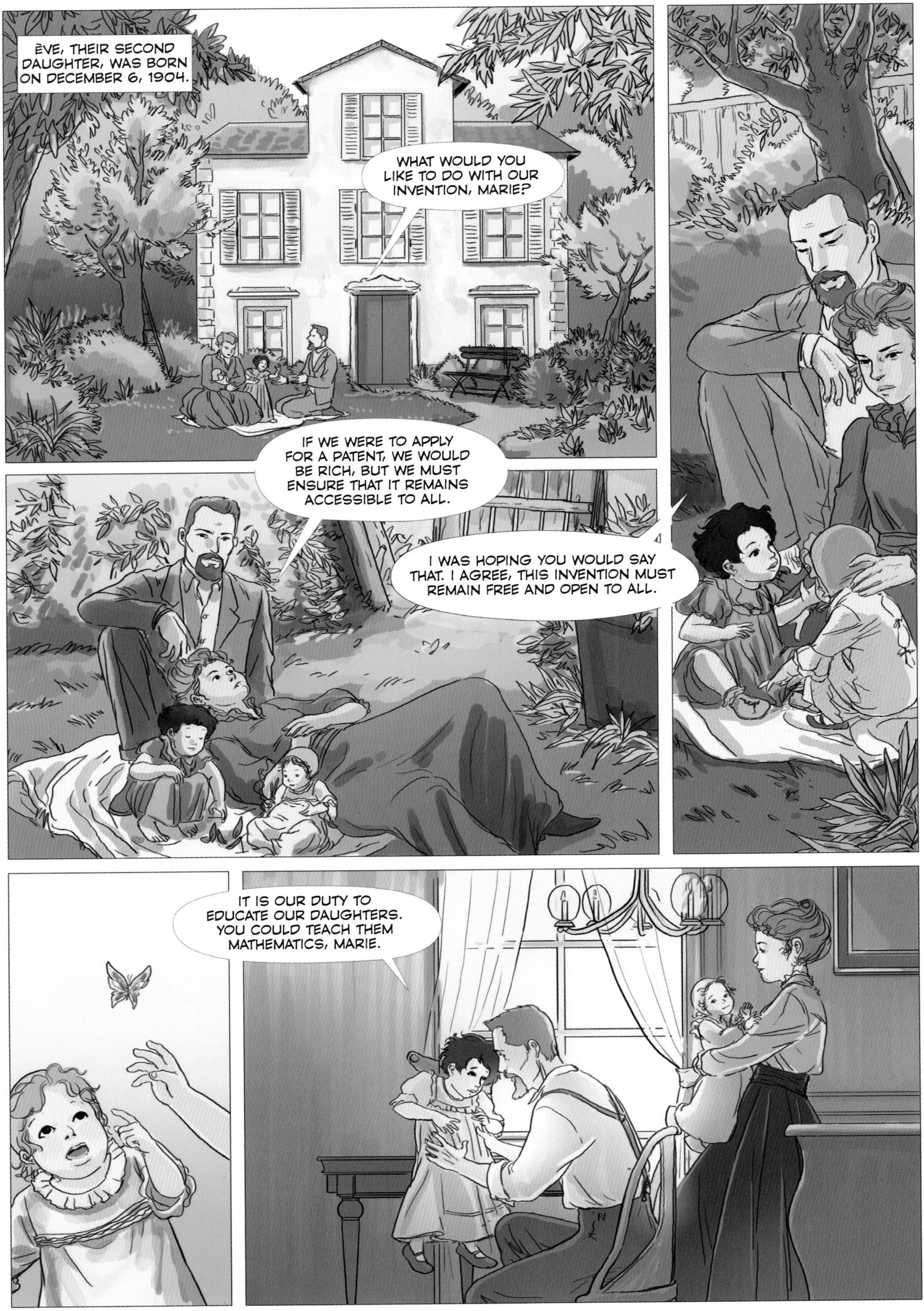
ÈVE, THEIR SECOND DAUGHTER, WAS BORN ON DECEMBER 6, 1904.
WHAT WOULD YOU LIKE TO DO WITH OUR INVENTION, MARIE?
IF WE WERE TO APPLY FOR A PATENT, WE WOULD BE RICH, BUT WE MUST ENSURE THAT IT REMAINS ACCESSIBLE TO ALL.
I WAS HOPING YOU WOULD SAY THAT. I AGREE, THIS INVENTION MUST REMAIN FREE AND OPEN TO ALL.
IT IS OUR DUTY TO EDUCATE OUR DAUGHTERS. YOU COULD TEACH THEM MATHEMATICS, MARIE.

SADLY, TRAGEDY STRUCK ON A RAINY APRIL 19, 1906, WHEN PIERRE CURIE WAS RUN OVER BY A CART ON A SLIPPERY STREET. HE DIED ON THE SPOT.

WE'RE HERE IF YOU NEED ANYTHING, MARIE...

MINISTER, WE MUST HELP MARIE CURIE, PIERRE CURIE'S WIDOW.

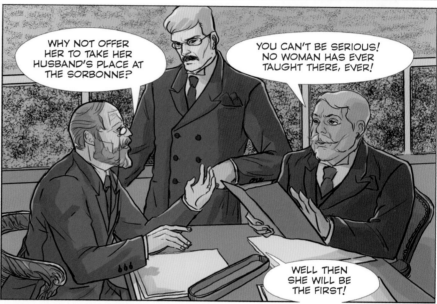

WHY NOT OFFER HER TO TAKE HER HUSBAND'S PLACE AT THE SORBONNE?

YOU CAN'T BE SERIOUS! NO WOMAN HAS EVER TAUGHT THERE, EVER!

WELL THEN SHE WILL BE THE FIRST!

"I'VE BEEN OFFERED TO TAKE OVER YOUR CLASS AND YOUR LAB, MY PIERRE. I ACCEPTED. I DON'T KNOW IF I DID THE RIGHT THING."

"BUT YOU DID OFTEN TELL ME YOU WISHED I COULD TEACH A CLASS AT THE SORBONNE. BESIDES, I WOULD LIKE TO MAKE SOME EFFORT TO PURSUE OUR RESEARCH..."

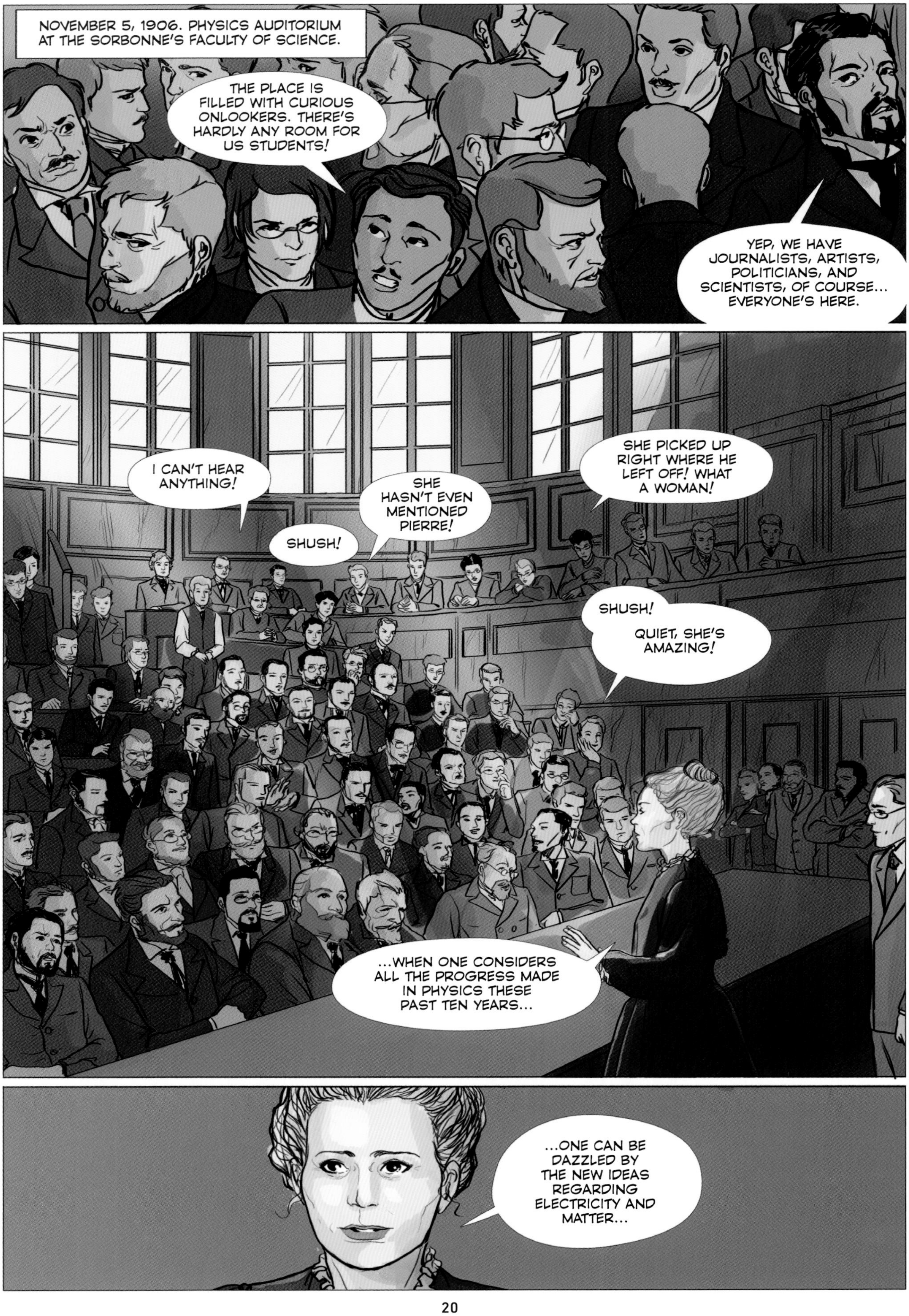
NOVEMBER 5, 1906. PHYSICS AUDITORIUM AT THE SORBONNE'S FACULTY OF SCIENCE.
THE PLACE IS FILLED WITH CURIOUS ONLOOKERS. THERE'S HARDLY ANY ROOM FOR US STUDENTS!
YEP, WE HAVE JOURNALISTS, ARTISTS, POLITICIANS, AND SCIENTISTS, OF COURSE... EVERYONE'S HERE.
I CAN'T HEAR ANYTHING!
SHUSH!
SHE HASN'T EVEN MENTIONED PIERRE!
SHE PICKED UP RIGHT WHERE HE LEFT OFF! WHAT A WOMAN!
SHUSH!
QUIET, SHE'S AMAZING!
...WHEN ONE CONSIDERS ALL THE PROGRESS MADE IN PHYSICS THESE PAST TEN YEARS...
...ONE CAN BE DAZZLED BY THE NEW IDEAS REGARDING ELECTRICITY AND MATTER...

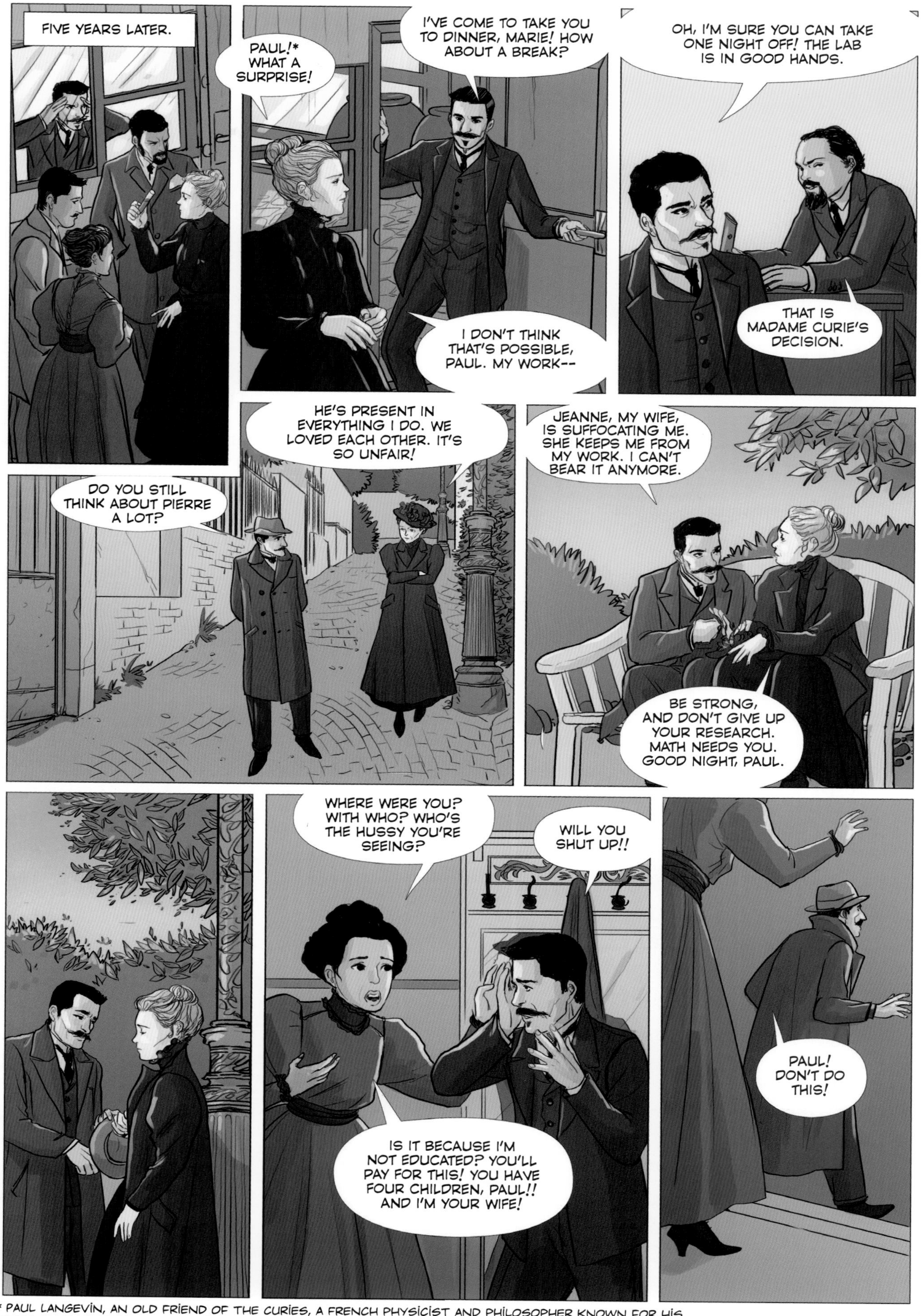

FIVE YEARS LATER.
PAUL!* WHAT A SURPRISE!
I'VE COME TO TAKE YOU TO DINNER, MARIE! HOW ABOUT A BREAK?
I DON'T THINK THAT'S POSSIBLE, PAUL. MY WORK--
OH, I'M SURE YOU CAN TAKE ONE NIGHT OFF! THE LAB IS IN GOOD HANDS.
THAT IS MADAME CURIE'S DECISION.
DO YOU STILL THINK ABOUT PIERRE A LOT?
HE'S PRESENT IN EVERYTHING I DO. WE LOVED EACH OTHER. IT'S SO UNFAIR!
JEANNE, MY WIFE, IS SUFFOCATING ME. SHE KEEPS ME FROM MY WORK. I CAN'T BEAR IT ANYMORE.
BE STRONG, AND DON'T GIVE UP YOUR RESEARCH. MATH NEEDS YOU. GOOD NIGHT, PAUL.
WHERE WERE YOU? WITH WHO? WHO'S THE HUSSY YOU'RE SEEING?
WILL YOU SHUT UP!!
IS IT BECAUSE I'M NOT EDUCATED? YOU'LL PAY FOR THIS! YOU HAVE FOUR CHILDREN, PAUL!! AND I'M YOUR WIFE!
PAUL! DON'T DO THIS!

* PAUL LANGEVIN, AN OLD FRIEND OF THE CURIES, A FRENCH PHYSICIST AND PHILOSOPHER KNOWN FOR HIS WORK ON MAGNETISM AND FOR INTRODUCING EINSTEIN'S WORK TO FRANCE.

DID I DO THE RIGHT THING, MARIE?

WHILE PAUL LANGEVIN AND MARIE CURIE ATTENDED THE SOLVAY CONGRESS IN BRUSSELS WITH THE WORLD'S BEST PHYSICIANS, MATHEMATICIANS AND CHEMISTS, THE SCANDAL WAS ALL OVER THE FRENCH PRESS.

Le Journal

4 novembre 1911 N°13??

Love affair between Madame Curie and Professor Langevin

Paul Langevin's mother-in-law claims that Madame Curie has kidnapped the husband of her daughter, a mother of four

LE PETIT JOURNAL

Romance in a laboratory in

NO, I HAVE NOT KIDNAPPED ANYONE. I CONSIDER ANY VIOLATION OF PRIVACY BY THE PRESS AN ABOMINATION.

THIS VIOLATION IS PARTICULARLY CRIMINAL WHEN IT CONCERNS PEOPLE WHO HAVE CLEARLY DEVOTED THEIR LIVES TO HIGHER PREOCCUPATIONS THAT PROMOTE THE PUBLIC INTEREST.

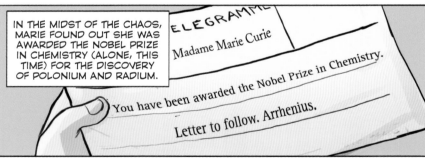

IN THE MIDST OF THE CHAOS, MARIE FOUND OUT SHE WAS AWARDED THE NOBEL PRIZE IN CHEMISTRY (ALONE, THIS TIME) FOR THE DISCOVERY OF POLONIUM AND RADIUM.

TELEGRAMME

Madame Marie Curie

You have been awarded the Nobel Prize in Chemistry.

Letter to follow. Arrhenius.

EXCELSIOR

Journal Illustré Quotidien

Les Scandales de la Sorbonne

GIL BLAS

Marie Curie, the Polish woman, Dreyfus defender and home wrecker in a 'France rife with immigra ...stroying and dishonori

L'INTRANSIGEA

23 NOVEMBRE

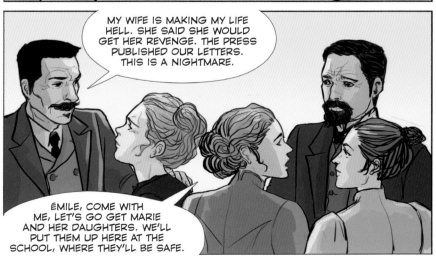

MY WIFE IS MAKING MY LIFE HELL. SHE SAID SHE WOULD GET HER REVENGE. THE PRESS PUBLISHED OUR LETTERS. THIS IS A NIGHTMARE.

ÉMILE, COME WITH ME, LET'S GO GET MARIE AND HER DAUGHTERS. WE'LL PUT THEM UP HERE AT THE SCHOOL, WHERE THEY'LL BE SAFE.

I HOPE IT GOES WELL.

HAVE FAITH IN YOUR MOTHER.

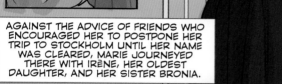

FIFTEEN YEARS AGO, THE RADIOACTIVITY OF URANIUM WAS DISCOVERED BY HENRI BECQUEREL, AND TWO YEARS LATER, THE STUDY OF THIS PHENOMENON WAS EXPANDED TO OTHER SUBSTANCES, FIRST BY MYSELF AND THEN BY PIERRE CURIE.

I CALLED ALL BODIES EMITTING SUCH ACTIVITY "RADIOACTIVE."

THE STORY OF THE DISCOVERY AND ISOLATION OF RADIUM PROVED THE THEORY I HAD ARTICULATED, WHEREBY RADIO-ACTIVITY IS AN ATOMIC PROPERTY OF MATTER AND CAN PROVIDE A METHOD OF RESEARCH WITH NEW ELEMENTS.

AGAINST THE ADVICE OF FRIENDS WHO ENCOURAGED HER TO POSTPONE HER TRIP TO STOCKHOLM UNTIL HER NAME WAS CLEARED, MARIE JOURNEYED THERE WITH IRÈNE, HER OLDEST DAUGHTER, AND HER SISTER BRONIA.

MÉ, I'M GLAD TO SEE YOU'RE FEELING BETTER.

I'M STILL QUITE TIRED, DARLING. I'LL GO GET SOME REST IN BRUNOY. YOU CAN WRITE ME THERE UNDER THE NAME SKLODOWSKA.

THE LAB WILL HAVE TO WAIT. I NEED TIME TO RECOVER FROM ALL THIS NEGATIVE PUBLICITY... AND FROM THIS SECOND NOBEL PRIZE!

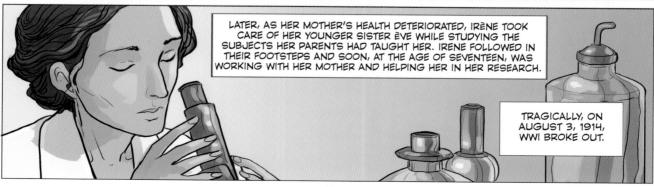

LATER, AS HER MOTHER'S HEALTH DETERIORATED, IRÈNE TOOK CARE OF HER YOUNGER SISTER ÈVE WHILE STUDYING THE SUBJECTS HER PARENTS HAD TAUGHT HER. IRENE FOLLOWED IN THEIR FOOTSTEPS AND SOON, AT THE AGE OF SEVENTEEN, WAS WORKING WITH HER MOTHER AND HELPING HER IN HER RESEARCH.

TRAGICALLY, ON AUGUST 3, 1914, WWI BROKE OUT.

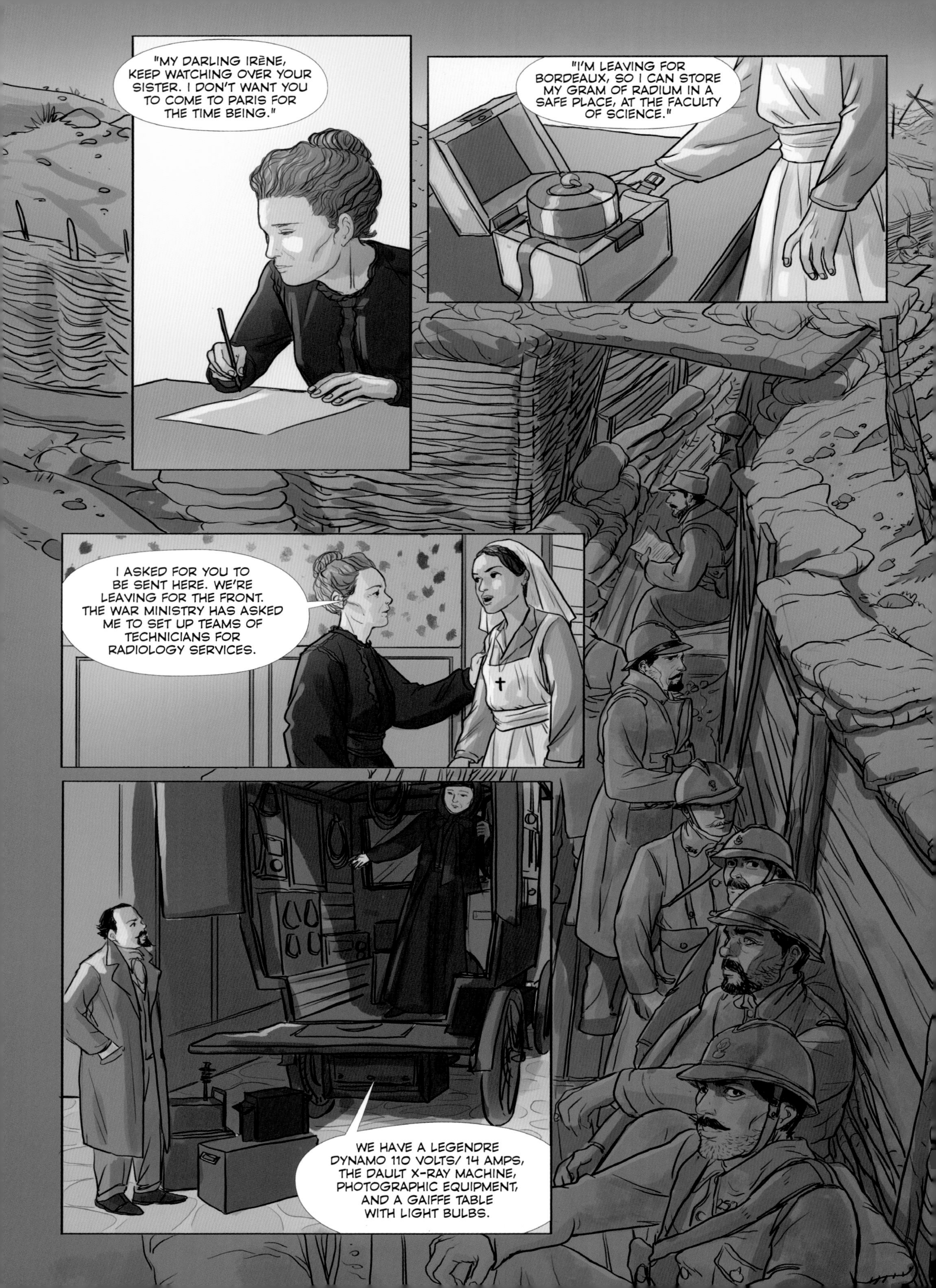
"MY DARLING IRÈNE, KEEP WATCHING OVER YOUR SISTER. I DON'T WANT YOU TO COME TO PARIS FOR THE TIME BEING."
"I'M LEAVING FOR BORDEAUX, SO I CAN STORE MY GRAM OF RADIUM IN A SAFE PLACE, AT THE FACULTY OF SCIENCE."
I ASKED FOR YOU TO BE SENT HERE. WE'RE LEAVING FOR THE FRONT. THE WAR MINISTRY HAS ASKED ME TO SET UP TEAMS OF TECHNICIANS FOR RADIOLOGY SERVICES.
WE HAVE A LEGENDRE DYNAMO 110 VOLTS/ 14 AMPS, THE DAULT X-RAY MACHINE, PHOTOGRAPHIC EQUIPMENT, AND A GAIFFE TABLE WITH LIGHT BULBS.

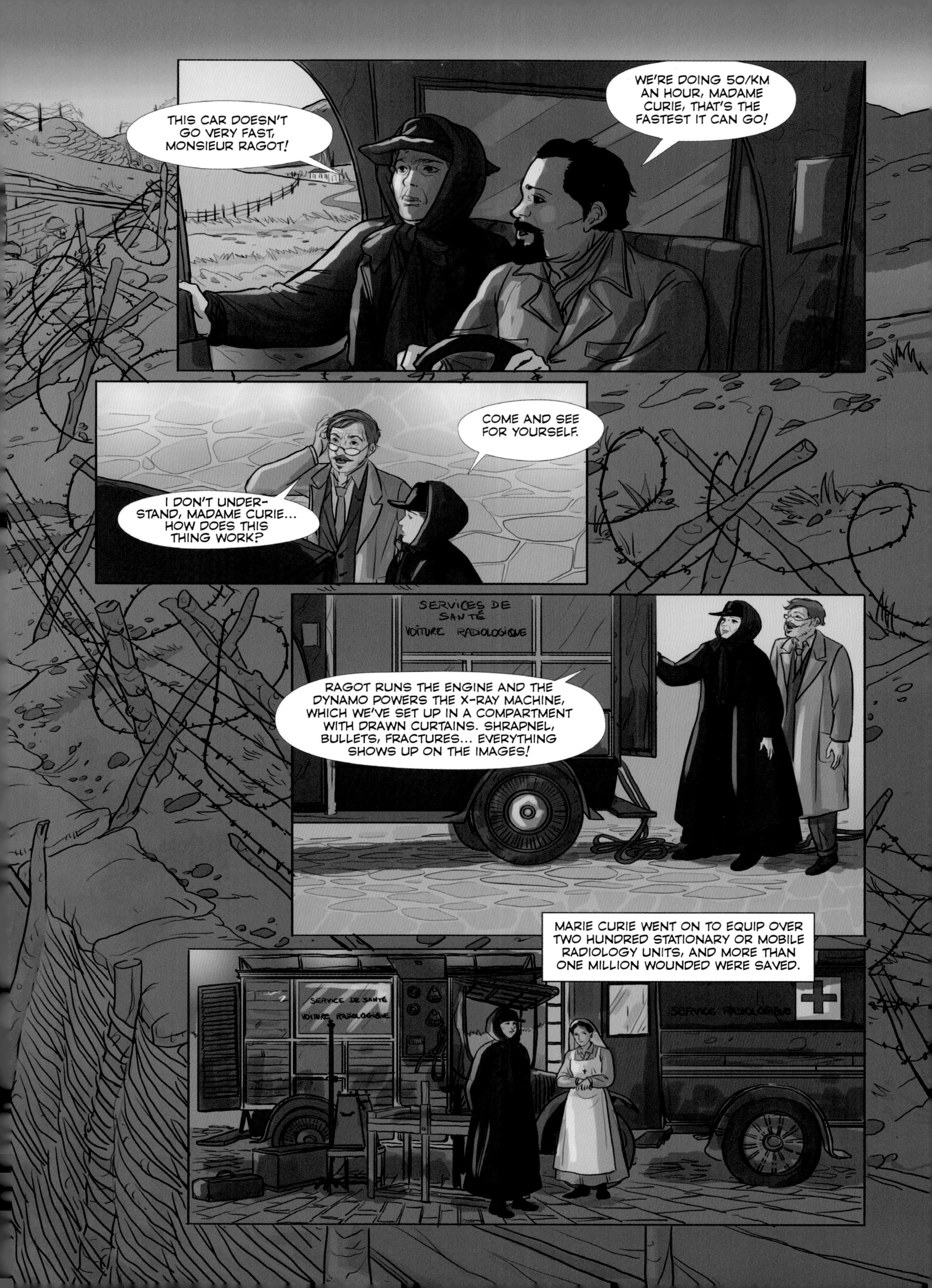

THIS CAR DOESN'T GO VERY FAST, MONSIEUR RAGOT!
WE'RE DOING 50/KM AN HOUR, MADAME CURIE, THAT'S THE FASTEST IT CAN GO!
I DON'T UNDER-STAND, MADAME CURIE... HOW DOES THIS THING WORK?
COME AND SEE FOR YOURSELF.
SERVICES DE SANTÉ
VOITURE RADIOLOGIQUE
RAGOT RUNS THE ENGINE AND THE DYNAMO POWERS THE X-RAY MACHINE, WHICH WE'VE SET UP IN A COMPARTMENT WITH DRAWN CURTAINS. SHRAPNEL, BULLETS, FRACTURES... EVERYTHING SHOWS UP ON THE IMAGES!
MARIE CURIE WENT ON TO EQUIP OVER TWO HUNDRED STATIONARY OR MOBILE RADIOLOGY UNITS, AND MORE THAN ONE MILLION WOUNDED WERE SAVED.
SERVICE DE SANTÉ
VOITURE RADIOLOGIQUE
SERVICE RADIOLOGIQUE

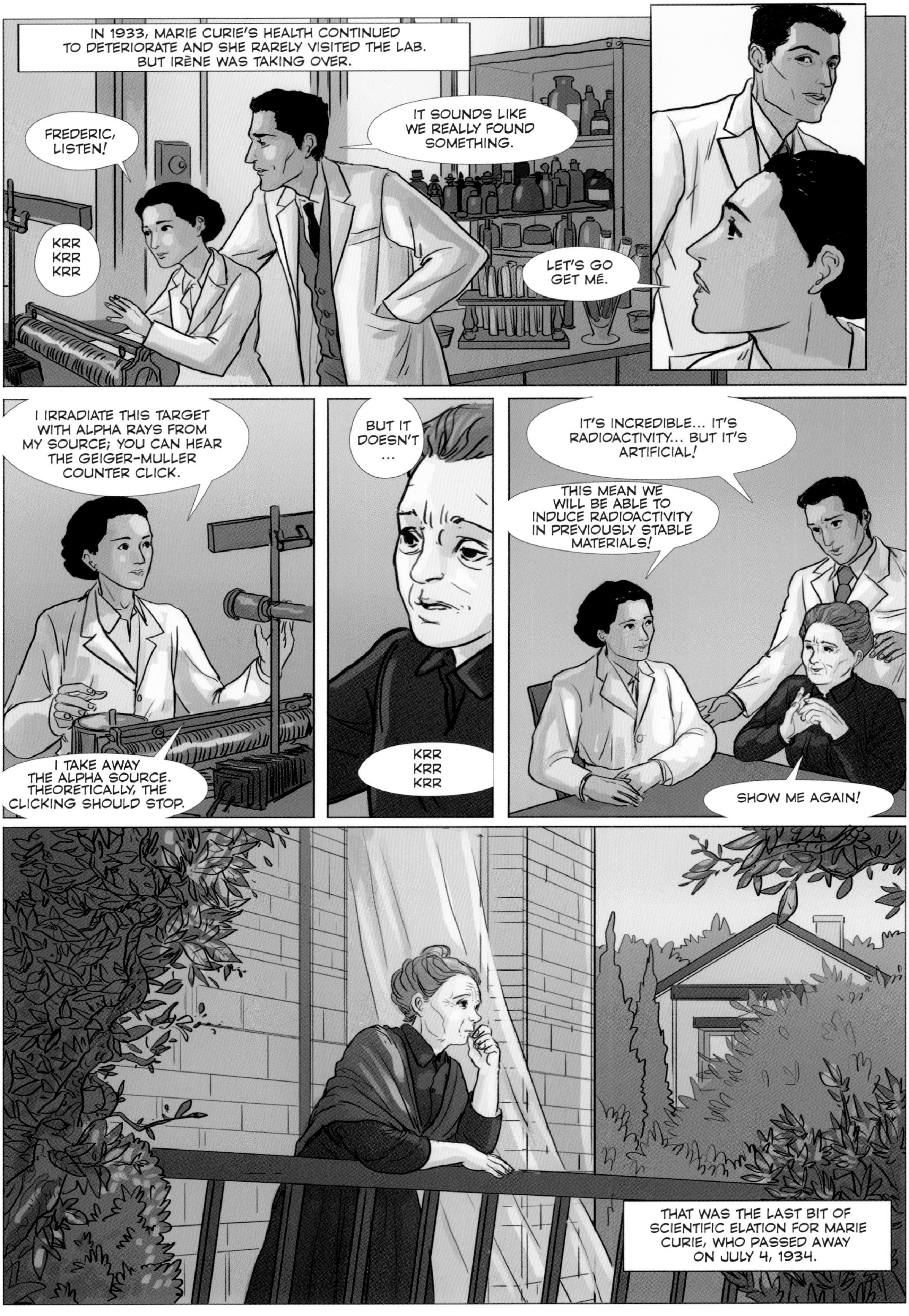
IN 1933, MARIE CURIE'S HEALTH CONTINUED TO DETERIORATE AND SHE RARELY VISITED THE LAB. BUT IRÈNE WAS TAKING OVER.
FREDERIC, LISTEN!
KRR KRR KRR
IT SOUNDS LIKE WE REALLY FOUND SOMETHING.
LET'S GO GET MÉ.
I IRRADIATE THIS TARGET WITH ALPHA RAYS FROM MY SOURCE; YOU CAN HEAR THE GEIGER-MULLER COUNTER CLICK.
I TAKE AWAY THE ALPHA SOURCE. THEORETICALLY, THE CLICKING SHOULD STOP.
BUT IT DOESN'T ...
KRR KRR KRR
IT'S INCREDIBLE... IT'S RADIOACTIVITY... BUT IT'S ARTIFICIAL!
THIS MEAN WE WILL BE ABLE TO INDUCE RADIOACTIVITY IN PREVIOUSLY STABLE MATERIALS!
SHOW ME AGAIN!
THAT WAS THE LAST BIT OF SCIENTIFIC ELATION FOR MARIE CURIE, WHO PASSED AWAY ON JULY 4, 1934.

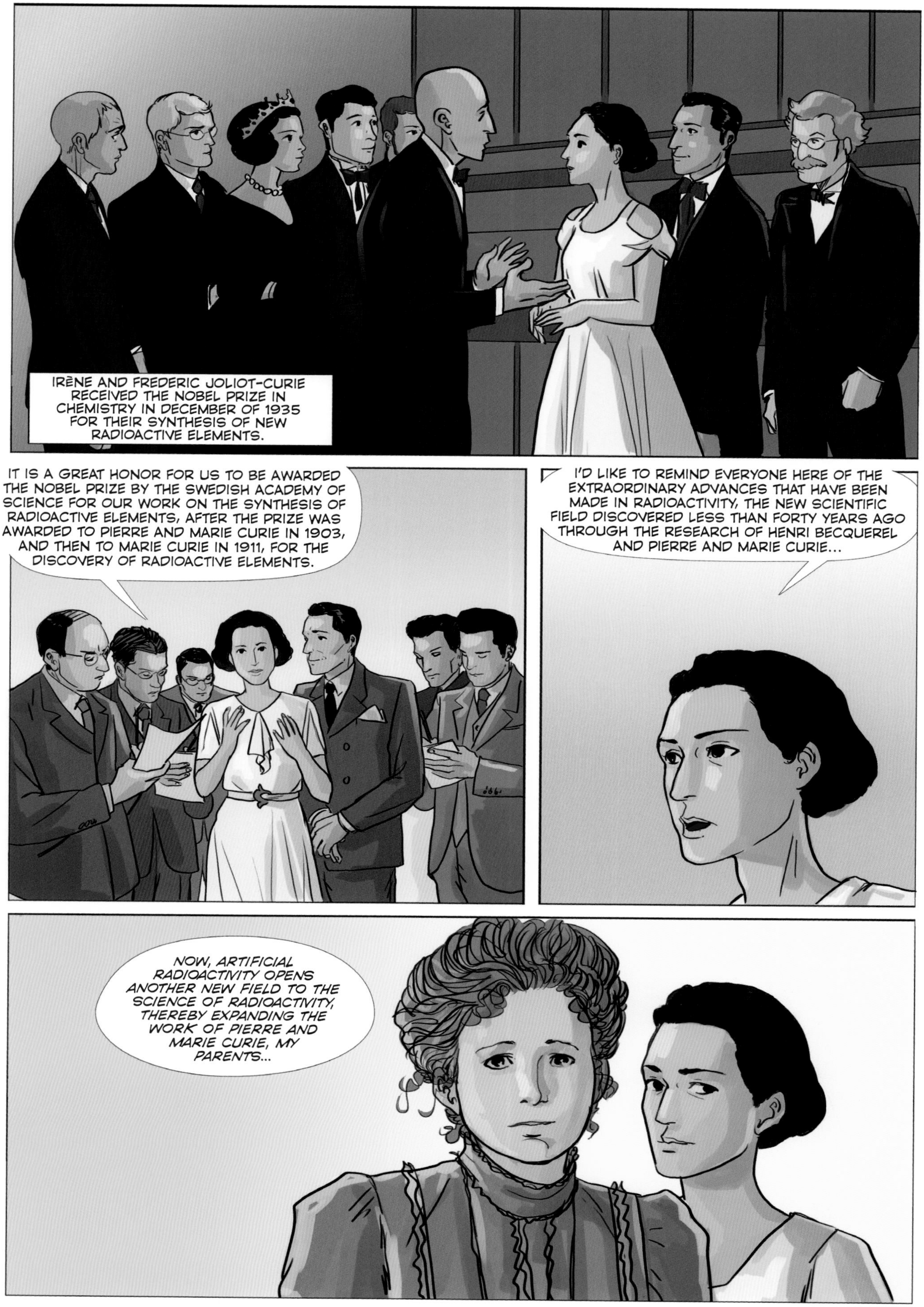
IRÈNE AND FREDERIC JOLIOT-CURIE RECEIVED THE NOBEL PRIZE IN CHEMISTRY IN DECEMBER OF 1935 FOR THEIR SYNTHESIS OF NEW RADIOACTIVE ELEMENTS.
IT IS A GREAT HONOR FOR US TO BE AWARDED THE NOBEL PRIZE BY THE SWEDISH ACADEMY OF SCIENCE FOR OUR WORK ON THE SYNTHESIS OF RADIOACTIVE ELEMENTS, AFTER THE PRIZE WAS AWARDED TO PIERRE AND MARIE CURIE IN 1903, AND THEN TO MARIE CURIE IN 1911, FOR THE DISCOVERY OF RADIOACTIVE ELEMENTS.
I'D LIKE TO REMIND EVERYONE HERE OF THE EXTRAORDINARY ADVANCES THAT HAVE BEEN MADE IN RADIOACTIVITY, THE NEW SCIENTIFIC FIELD DISCOVERED LESS THAN FORTY YEARS AGO THROUGH THE RESEARCH OF HENRI BECQUEREL AND PIERRE AND MARIE CURIE...
NOW, ARTIFICIAL RADIOACTIVITY OPENS ANOTHER NEW FIELD TO THE SCIENCE OF RADIOACTIVITY, THEREBY EXPANDING THE WORK OF PIERRE AND MARIE CURIE, MY PARENTS...

THE END

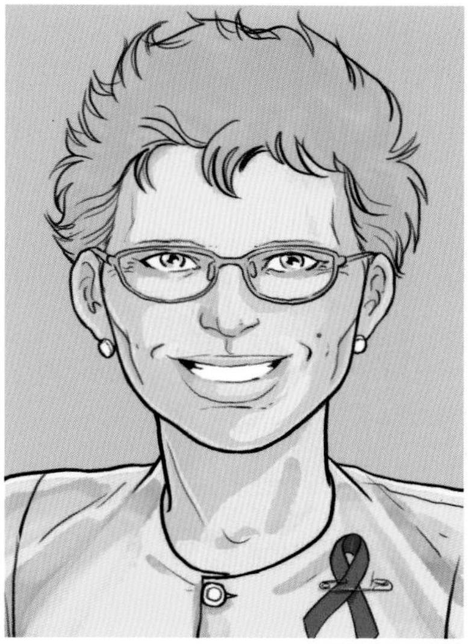

Françoise Barré-Sinoussi

1947 -

French virologist
Nobel Prize in Medicine 2008

An only child fascinated by all living things, **Françoise Barré-Sinoussi** spent her childhood observing the teeming life in the meadows of Auvergne.

Born into a family of modest means, she would have liked to study medicine but ultimately chose the sciences, primarily for economic reasons. She was a brilliant student, which paved her way to the United States and opened the door to prestigious laboratories, including the main one, that of Jean-Claude Chermann at the Pasteur Institute.

It was there that, in 1983, as part of Professor Montagnier's team, that she discovered HIV, the retrovirus responsible for AIDS.

In 2008, she and Luc Montagnier were awarded the Nobel Prize in medicine for this discovery.

A former research director at Inserm, professor at the Institut Pasteur until 2015, president of the International AIDS Society (IAS) from 2012 to 2014, she is now a member of the French National Academy of Sciences and president of Sidaction, a French fundraising society to fight AIDS..

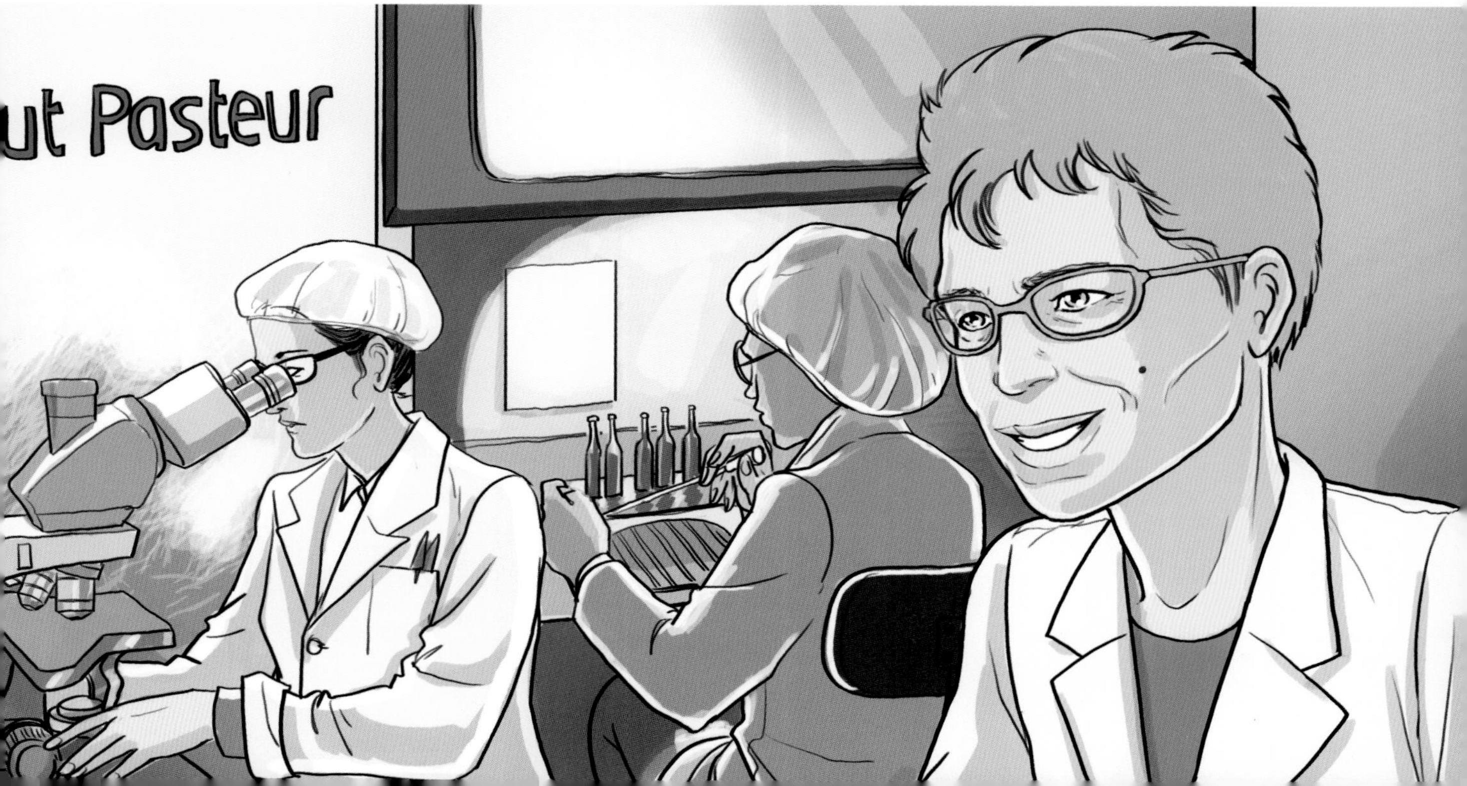
ut Pasteur

Donna Theo Strickland
1959 –

Canadian physicist
Nobel Prize in Physics 2018
(shared with Gérard Mourou)

Born in 1959 in Guelph, Canada, **Donna Theo Strickland** is a pioneering Canadian physicist in the field of lasers.

In 1982, while she was still a student, she and Professor Gerard Mourou developed a technique for amplifying pulses with lasers, which was capable of producing ultrashort pulses of high intensity without damaging the equipment. This research became the subject of a publication.

It was on the basis of that first article that she received the 2018 Nobel Prize in Physics, which she shared with Gérard Mourou, for advances in laser physics (laser technology itself was created in 1960).

She thus became the third woman to be awarded the Nobel Laureate in physics, following Marie Curie in 1903 and Maria Goeppert-Mayer in 1963.

Strickland, who was largely unknown until her Nobel win, is now a professor in the physics and astronomy department at the University of Waterloo.

Laser technique is used all over the world, especially in eye surgery.

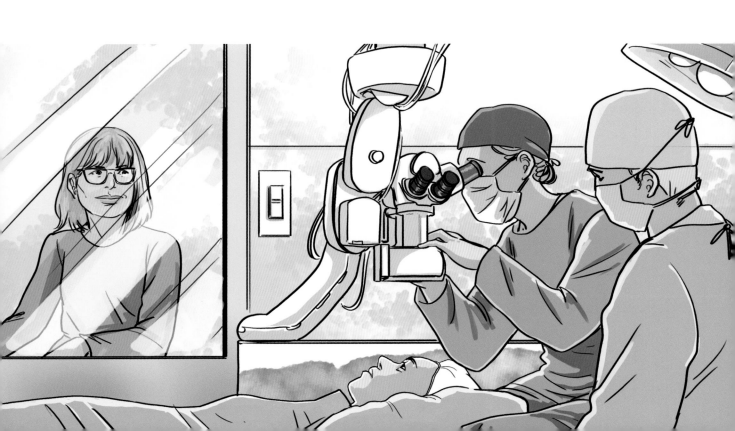

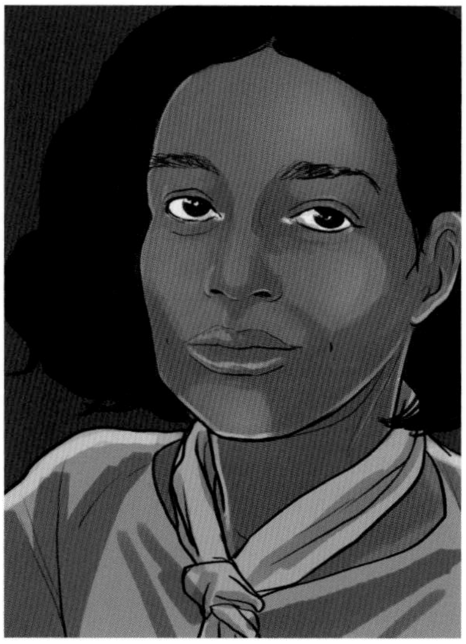

Dorothy Vaughan
1910 - 2008

American mathematician and computer scientist
Congressional Gold Medal

Dorothy Johnson Vaughan was born in Kansas City, Missouri in 1910. It was clear early on that she was gifted, as she successfully completed various academic courses at segregated schools in West Virginia, where her family had moved.

Torn between her family life (she was the mother of three daughters) and her teaching career in a segregated school, she ended up responding to the recruitment announcement of the NACA (which later become NASA), at the same time as Katherine Johnson (see p.54).

There, she joined the team of "computers in skirts" and went on to become the first black manager in the history of NASA when she was chosen to head up the west wing of the computer unit.

Aware of the rapid evolution of the potential of computer technology, she trained in the Fortran programming language, which enabled her to work on flight calculations with the SCOUT launch vehicle program (Solid Controlled Orbital Utility Test system), one of the most reliable launchers.

She was awarded the Congressional Gold Medal posthumously.

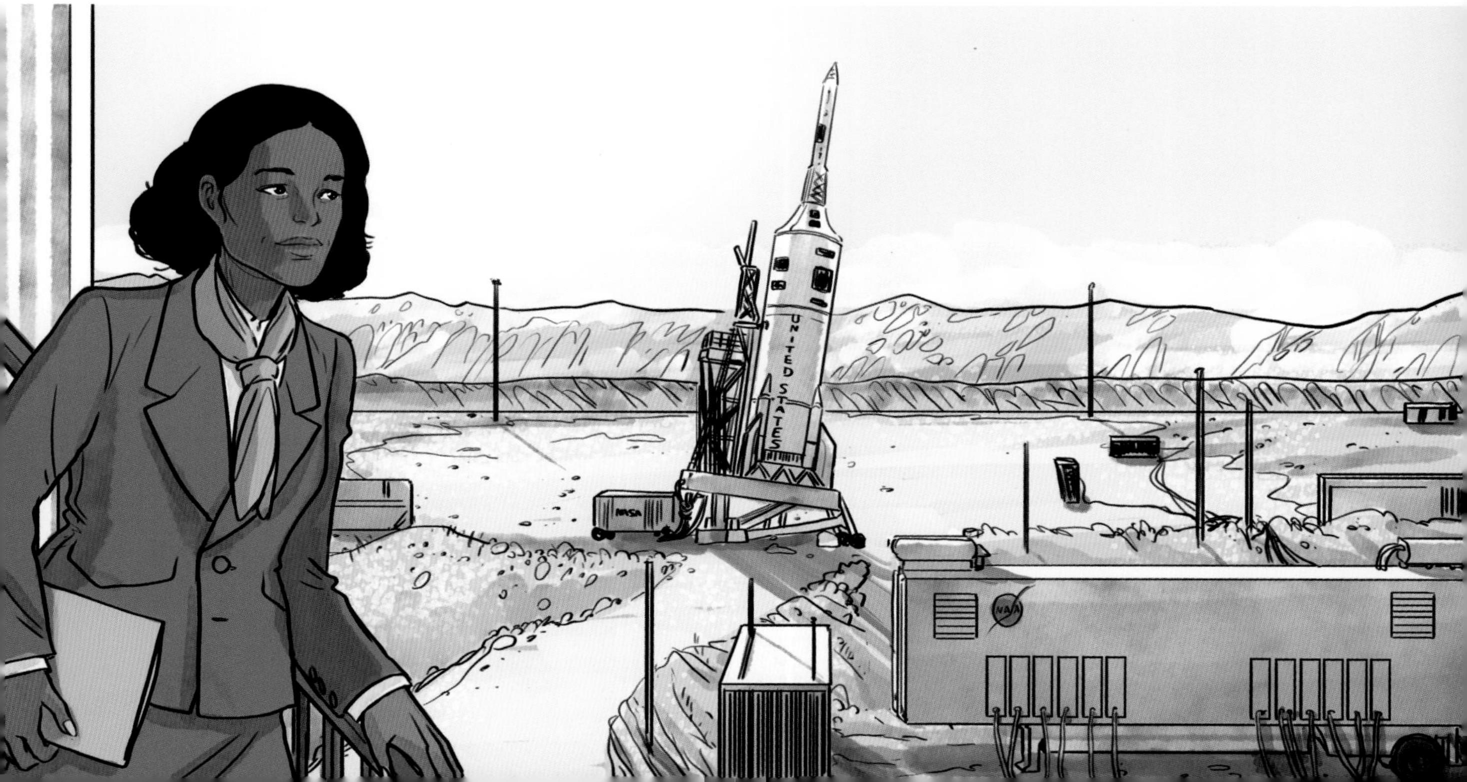
UNITED STATES
NASA

ADA LOVELACE

1815 - 1852
English; the inventor of computer programming

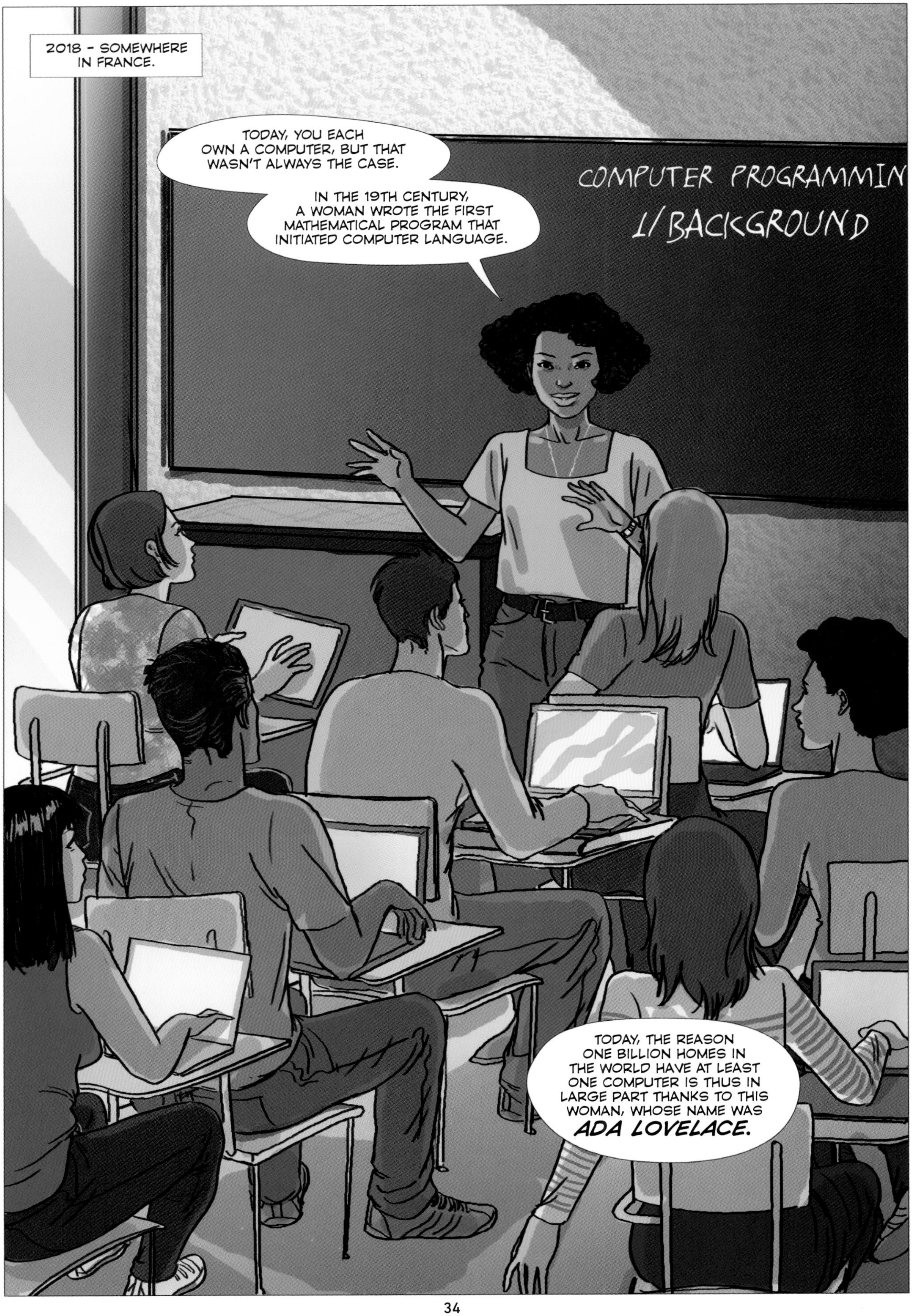
2018 - SOMEWHERE IN FRANCE.
TODAY, YOU EACH OWN A COMPUTER, BUT THAT WASN'T ALWAYS THE CASE.
IN THE 19TH CENTURY, A WOMAN WROTE THE FIRST MATHEMATICAL PROGRAM THAT INITIATED COMPUTER LANGUAGE.
COMPUTER PROGRAMMIN
1/ BACKGROUND
TODAY, THE REASON ONE BILLION HOMES IN THE WORLD HAVE AT LEAST ONE COMPUTER IS THUS IN LARGE PART THANKS TO THIS WOMAN, WHOSE NAME WAS ***ADA LOVELACE.***

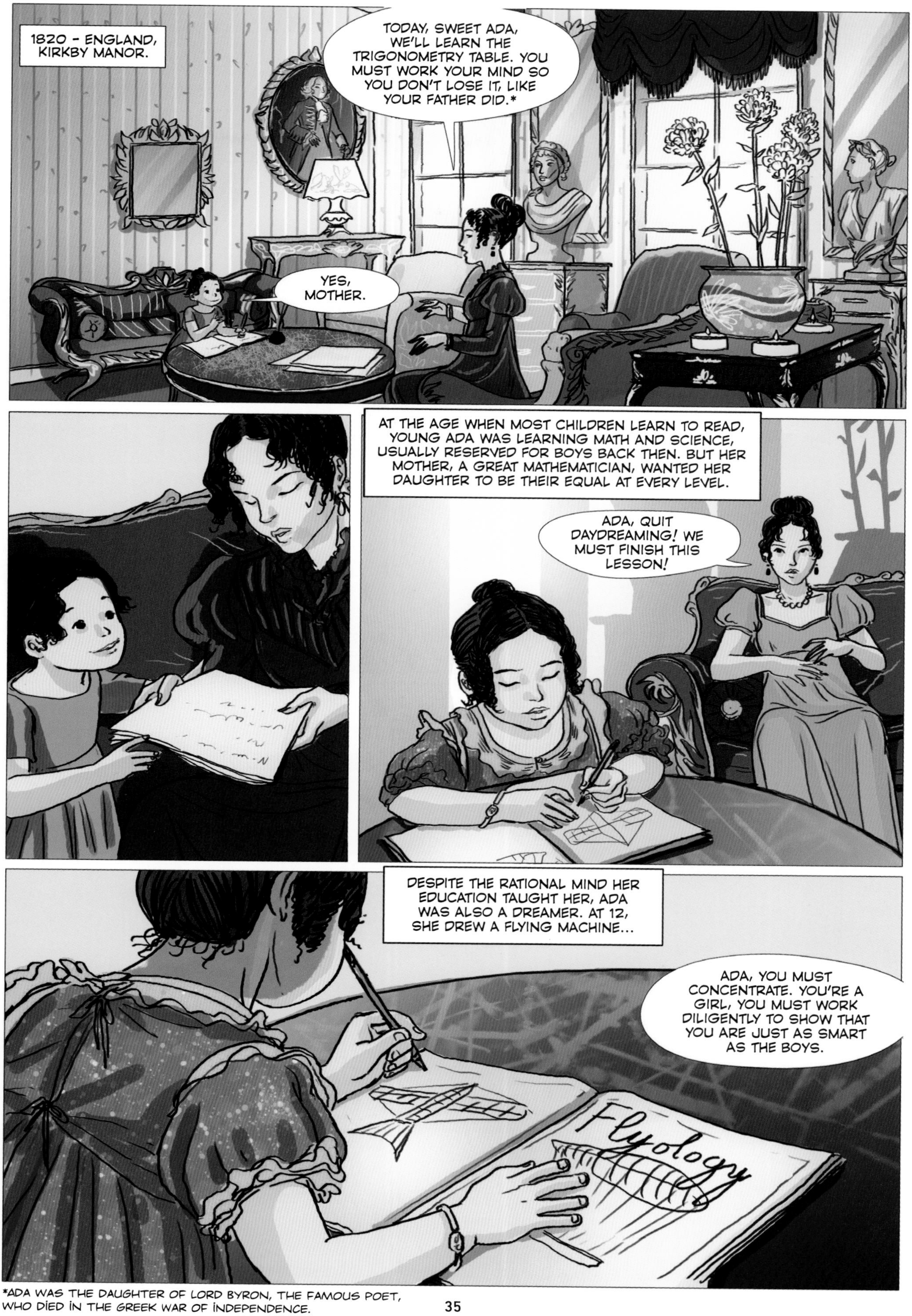
1820 - ENGLAND, KIRKBY MANOR.
TODAY, SWEET ADA, WE'LL LEARN THE TRIGONOMETRY TABLE. YOU MUST WORK YOUR MIND SO YOU DON'T LOSE IT, LIKE YOUR FATHER DID.*
YES, MOTHER.
AT THE AGE WHEN MOST CHILDREN LEARN TO READ, YOUNG ADA WAS LEARNING MATH AND SCIENCE, USUALLY RESERVED FOR BOYS BACK THEN. BUT HER MOTHER, A GREAT MATHEMATICIAN, WANTED HER DAUGHTER TO BE THEIR EQUAL AT EVERY LEVEL.
ADA, QUIT DAYDREAMING! WE MUST FINISH THIS LESSON!
DESPITE THE RATIONAL MIND HER EDUCATION TAUGHT HER, ADA WAS ALSO A DREAMER. AT 12, SHE DREW A FLYING MACHINE...
ADA, YOU MUST CONCENTRATE. YOU'RE A GIRL, YOU MUST WORK DILIGENTLY TO SHOW THAT YOU ARE JUST AS SMART AS THE BOYS.
Flyology

*ADA WAS THE DAUGHTER OF LORD BYRON, THE FAMOUS POET, WHO DIED IN THE GREEK WAR OF INDEPENDENCE.

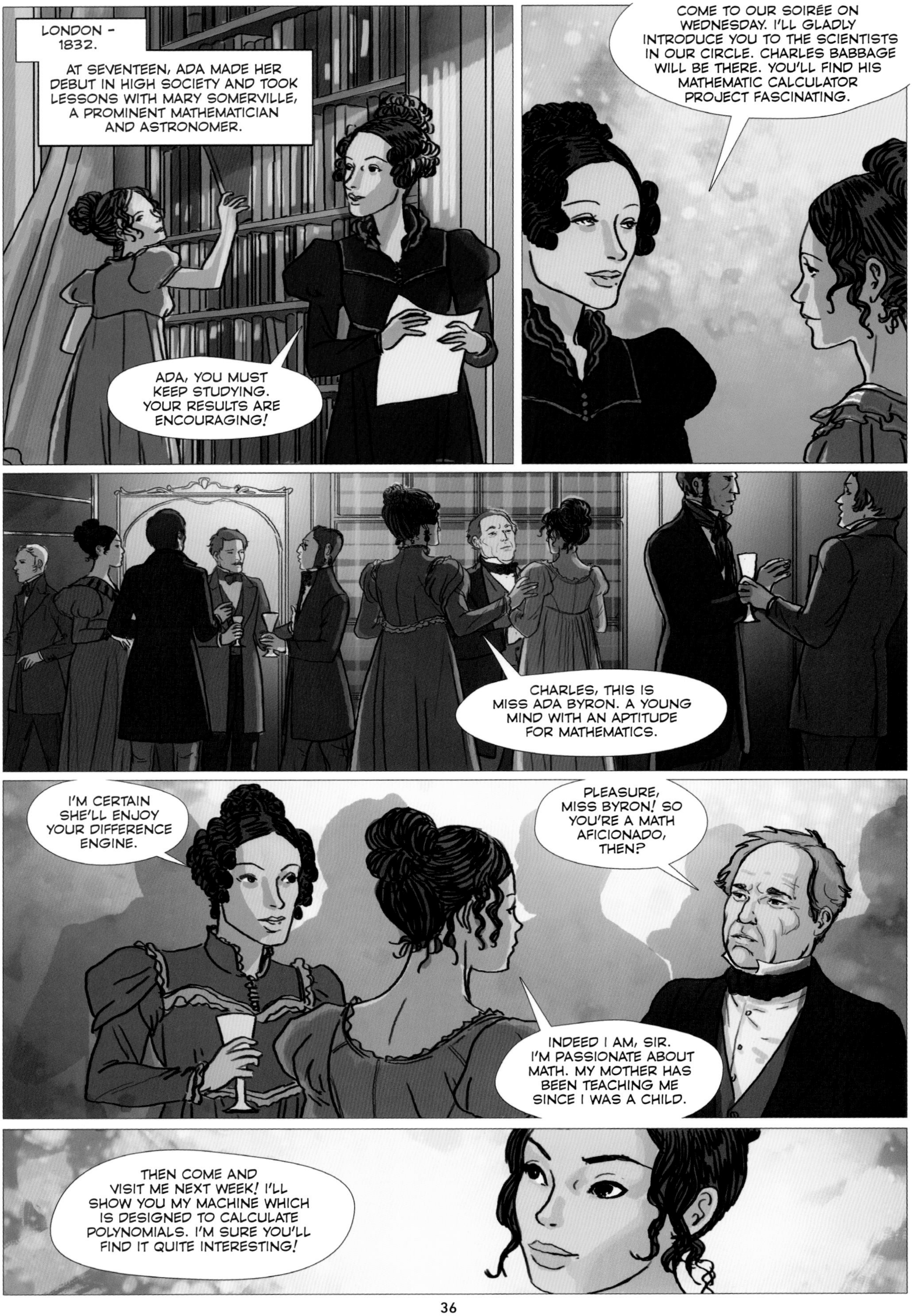
LONDON - 1832.
AT SEVENTEEN, ADA MADE HER DEBUT IN HIGH SOCIETY AND TOOK LESSONS WITH MARY SOMERVILLE, A PROMINENT MATHEMATICIAN AND ASTRONOMER.
ADA, YOU MUST KEEP STUDYING. YOUR RESULTS ARE ENCOURAGING!
COME TO OUR SOIRÉE ON WEDNESDAY. I'LL GLADLY INTRODUCE YOU TO THE SCIENTISTS IN OUR CIRCLE. CHARLES BABBAGE WILL BE THERE. YOU'LL FIND HIS MATHEMATIC CALCULATOR PROJECT FASCINATING.
CHARLES, THIS IS MISS ADA BYRON. A YOUNG MIND WITH AN APTITUDE FOR MATHEMATICS.
I'M CERTAIN SHE'LL ENJOY YOUR DIFFERENCE ENGINE.
PLEASURE, MISS BYRON! SO YOU'RE A MATH AFICIONADO, THEN?
INDEED I AM, SIR. I'M PASSIONATE ABOUT MATH. MY MOTHER HAS BEEN TEACHING ME SINCE I WAS A CHILD.
THEN COME AND VISIT ME NEXT WEEK! I'LL SHOW YOU MY MACHINE WHICH IS DESIGNED TO CALCULATE POLYNOMIALS. I'M SURE YOU'LL FIND IT QUITE INTERESTING!

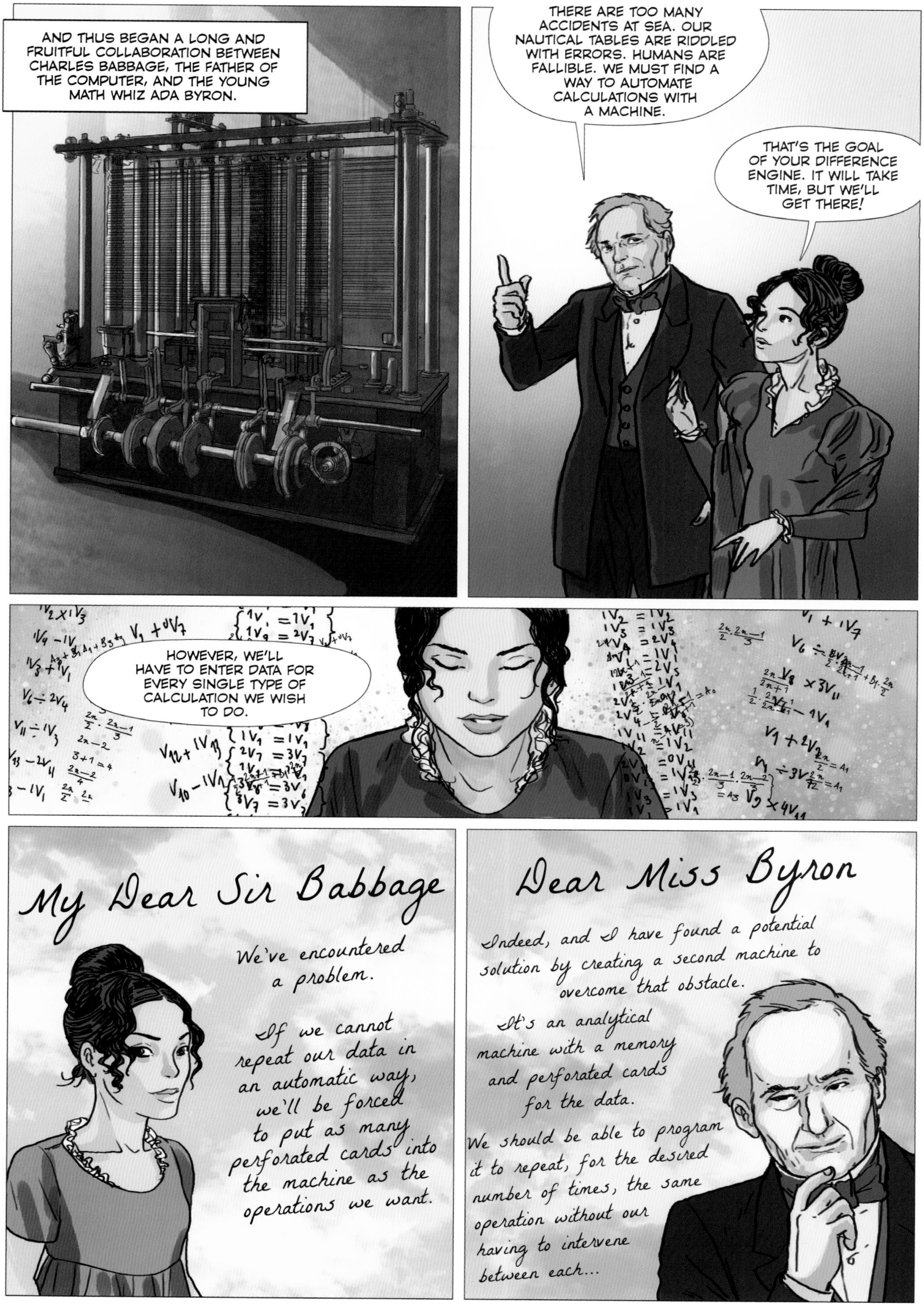
AND THUS BEGAN A LONG AND FRUITFUL COLLABORATION BETWEEN CHARLES BABBAGE, THE FATHER OF THE COMPUTER, AND THE YOUNG MATH WHIZ ADA BYRON.
THERE ARE TOO MANY ACCIDENTS AT SEA. OUR NAUTICAL TABLES ARE RIDDLED WITH ERRORS. HUMANS ARE FALLIBLE. WE MUST FIND A WAY TO AUTOMATE CALCULATIONS WITH A MACHINE.
THAT'S THE GOAL OF YOUR DIFFERENCE ENGINE. IT WILL TAKE TIME, BUT WE'LL GET THERE!
HOWEVER, WE'LL HAVE TO ENTER DATA FOR EVERY SINGLE TYPE OF CALCULATION WE WISH TO DO.
My Dear Sir Babbage
We've encountered a problem.
If we cannot repeat our data in an automatic way, we'll be forced to put as many perforated cards into the machine as the operations we want.
Dear Miss Byron
Indeed, and I have found a potential solution by creating a second machine to overcome that obstacle.
It's an analytical machine with a memory and perforated cards for the data.
We should be able to program it to repeat, for the desired number of times, the same operation without our having to intervene between each...

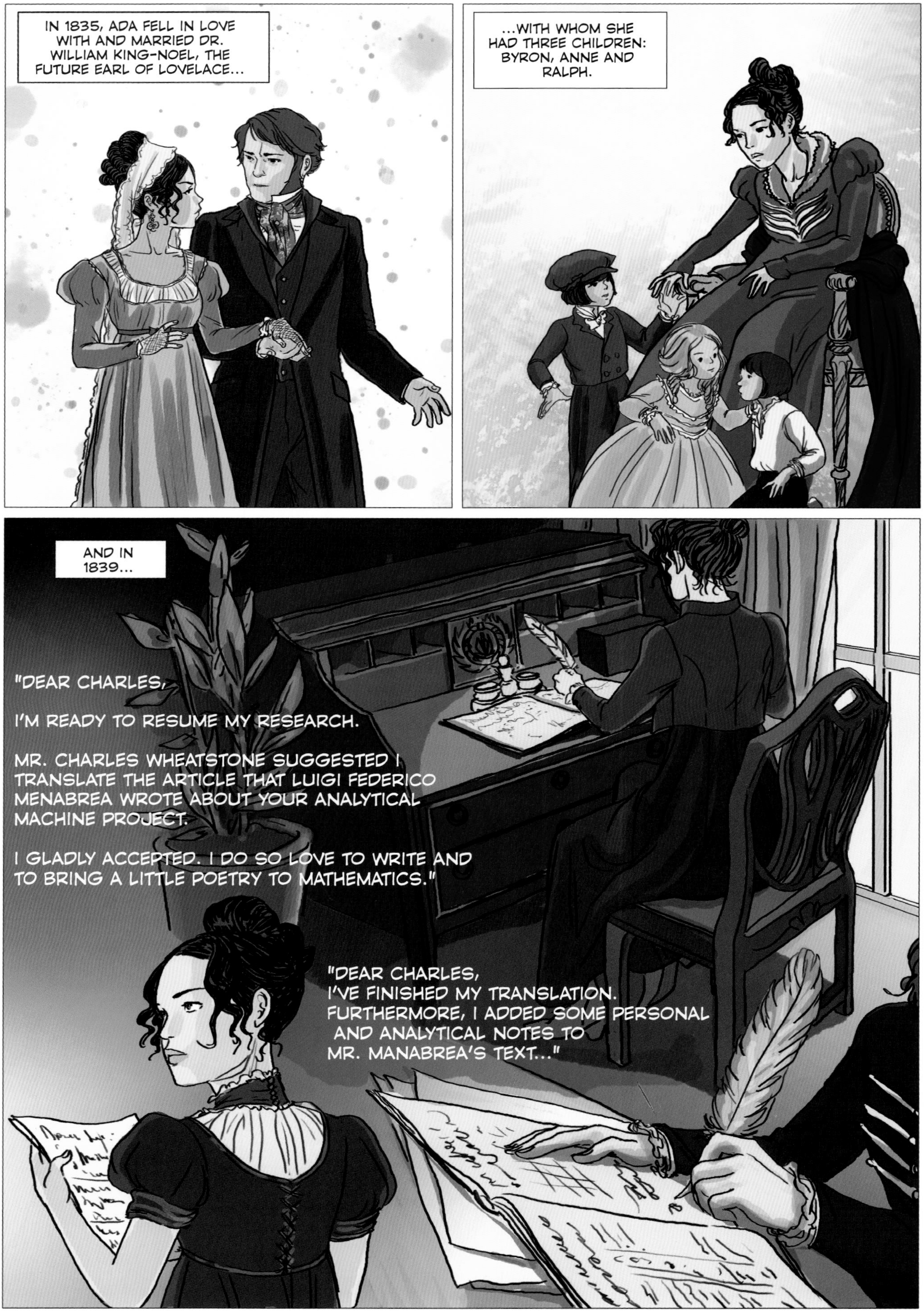
IN 1835, ADA FELL IN LOVE WITH AND MARRIED DR. WILLIAM KING-NOEL, THE FUTURE EARL OF LOVELACE...
...WITH WHOM SHE HAD THREE CHILDREN: BYRON, ANNE AND RALPH.
AND IN 1839...
"DEAR CHARLES,
I'M READY TO RESUME MY RESEARCH.
MR. CHARLES WHEATSTONE SUGGESTED I TRANSLATE THE ARTICLE THAT LUIGI FEDERICO MENABREA WROTE ABOUT YOUR ANALYTICAL MACHINE PROJECT.
I GLADLY ACCEPTED. I DO SO LOVE TO WRITE AND TO BRING A LITTLE POETRY TO MATHEMATICS."
"DEAR CHARLES,
I'VE FINISHED MY TRANSLATION.
FURTHERMORE, I ADDED SOME PERSONAL AND ANALYTICAL NOTES TO MR. MANABREA'S TEXT..."

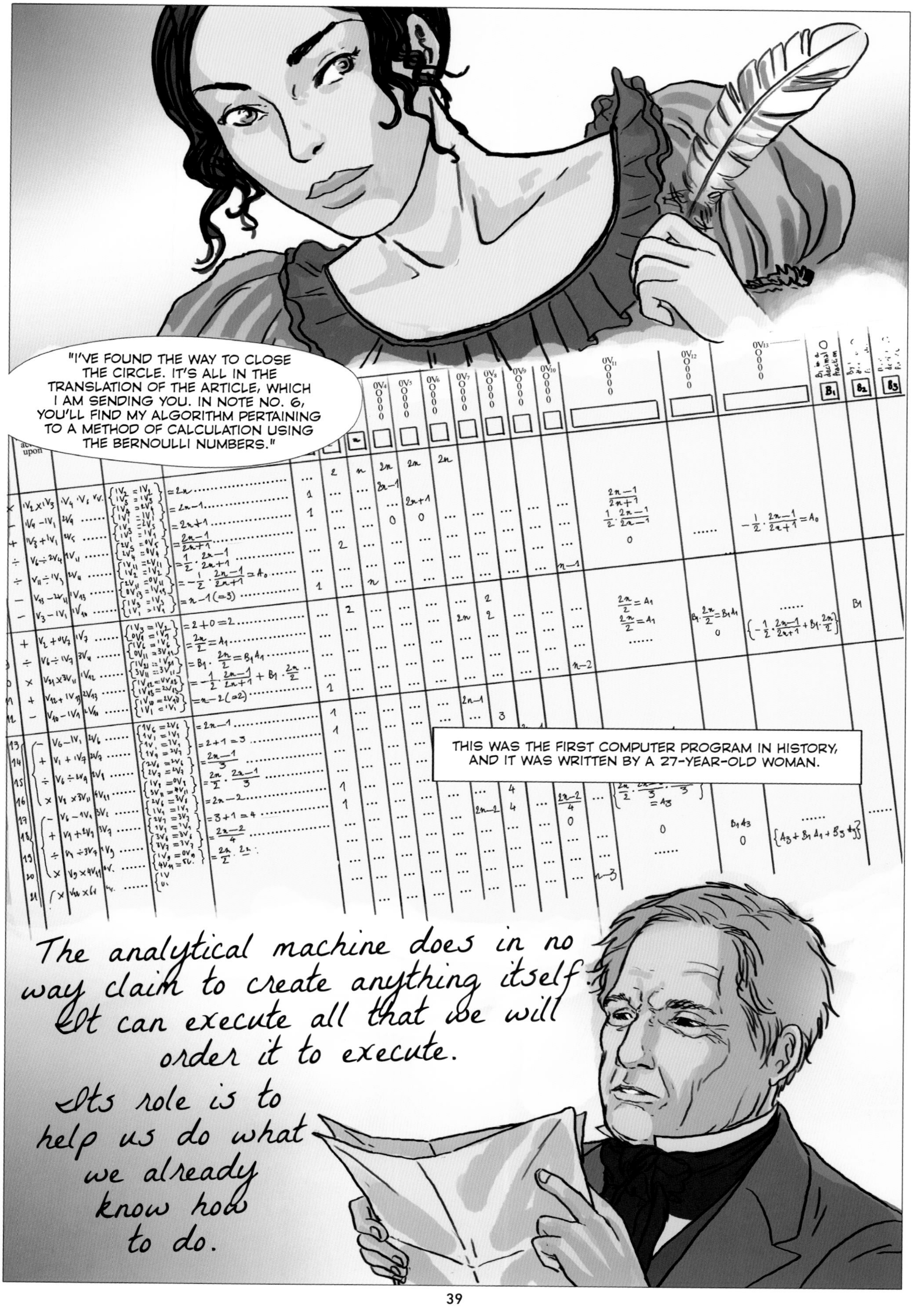
"I'VE FOUND THE WAY TO CLOSE THE CIRCLE. IT'S ALL IN THE TRANSLATION OF THE ARTICLE, WHICH I AM SENDING YOU. IN NOTE NO. 6, YOU'LL FIND MY ALGORITHM PERTAINING TO A METHOD OF CALCULATION USING THE BERNOULLI NUMBERS."
THIS WAS THE FIRST COMPUTER PROGRAM IN HISTORY, AND IT WAS WRITTEN BY A 27-YEAR-OLD WOMAN.
The analytical machine does in no way claim to create anything itself. It can execute all that we will order it to execute.
Its role is to help us do what we already know how to do.

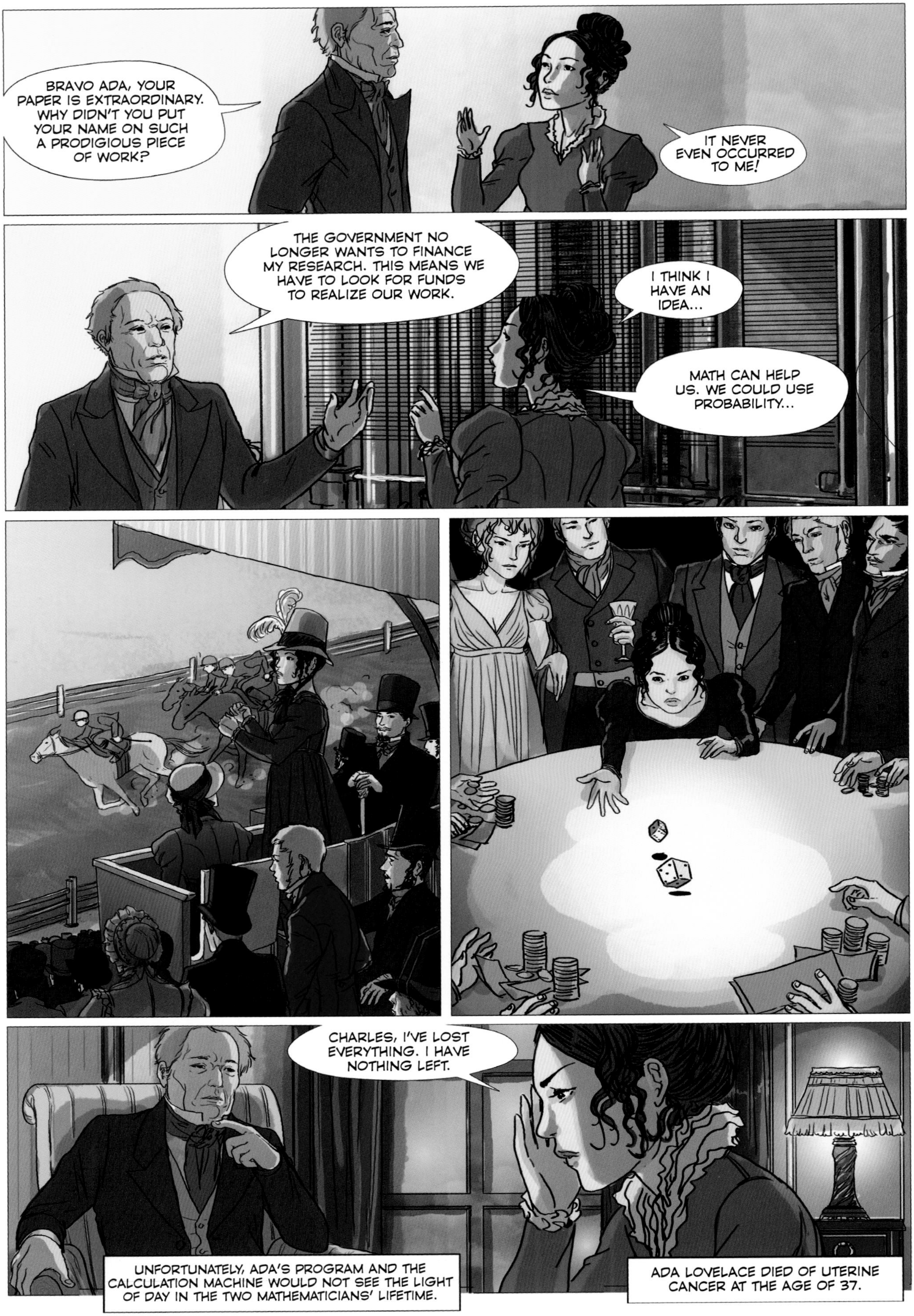
BRAVO ADA, YOUR PAPER IS EXTRAORDINARY. WHY DIDN'T YOU PUT YOUR NAME ON SUCH A PRODIGIOUS PIECE OF WORK?
IT NEVER EVEN OCCURRED TO ME!
THE GOVERNMENT NO LONGER WANTS TO FINANCE MY RESEARCH. THIS MEANS WE HAVE TO LOOK FOR FUNDS TO REALIZE OUR WORK.
I THINK I HAVE AN IDEA...
MATH CAN HELP US. WE COULD USE PROBABILITY...
CHARLES, I'VE LOST EVERYTHING. I HAVE NOTHING LEFT.
UNFORTUNATELY, ADA'S PROGRAM AND THE CALCULATION MACHINE WOULD NOT SEE THE LIGHT OF DAY IN THE TWO MATHEMATICIANS' LIFETIME.
ADA LOVELACE DIED OF UTERINE CANCER AT THE AGE OF 37.

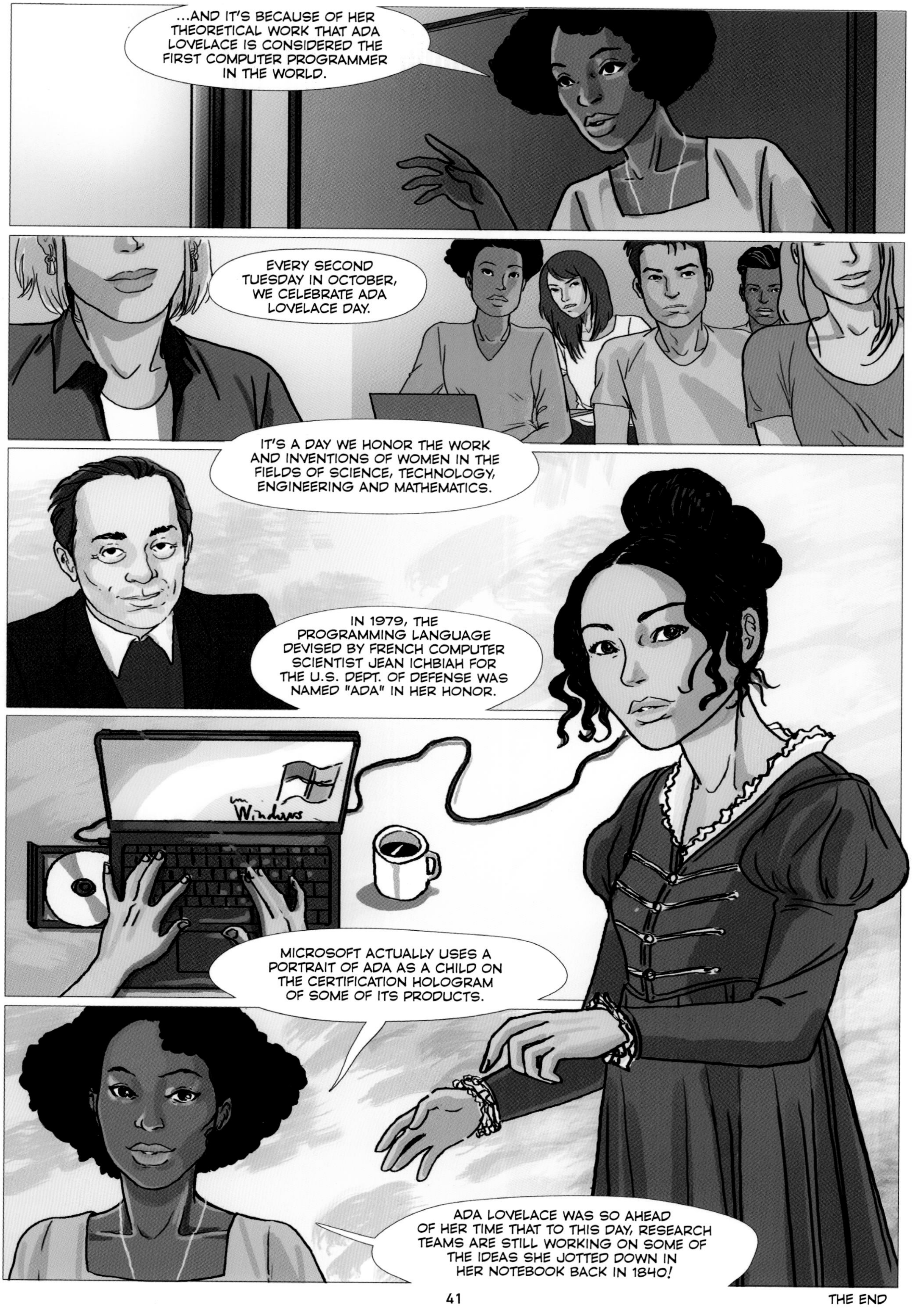
...AND IT'S BECAUSE OF HER THEORETICAL WORK THAT ADA LOVELACE IS CONSIDERED THE FIRST COMPUTER PROGRAMMER IN THE WORLD.
EVERY SECOND TUESDAY IN OCTOBER, WE CELEBRATE ADA LOVELACE DAY.
IT'S A DAY WE HONOR THE WORK AND INVENTIONS OF WOMEN IN THE FIELDS OF SCIENCE, TECHNOLOGY, ENGINEERING AND MATHEMATICS.
IN 1979, THE PROGRAMMING LANGUAGE DEVISED BY FRENCH COMPUTER SCIENTIST JEAN ICHBIAH FOR THE U.S. DEPT. OF DEFENSE WAS NAMED "ADA" IN HER HONOR.
Windows
MICROSOFT ACTUALLY USES A PORTRAIT OF ADA AS A CHILD ON THE CERTIFICATION HOLOGRAM OF SOME OF ITS PRODUCTS.
ADA LOVELACE WAS SO AHEAD OF HER TIME THAT TO THIS DAY, RESEARCH TEAMS ARE STILL WORKING ON SOME OF THE IDEAS SHE JOTTED DOWN IN HER NOTEBOOK BACK IN 1840!

THE END

Émilie du Châtelet
1706 - 1749

French mathematician and physicist

Gabrielle Emilie Le Tonnelier de Breteuil, Marquise du Chatelet, enjoyed a rich and intellectual education thanks to her father. He introduced her to the greatest scholars and philosophers of the time, which was quite unusual for a woman in those days.

A mathematics enthusiast, she translated Isaac Newton's Mathematical Principles of Natural Philosophy, which brought her immediate notoriety. Wishing to stimulate her mind in other ways besides the translation of complex formulas and principles of physics, she attended the discussions that the mathematician Maupertuis used to hold at the Café Gradot––for which she had to dress as a boy, as women were banned from cafés in those days.

Among her body of work is a treatise she wrote on physics, accompanied by experiments, which was published by the French Academy of Sciences, whose access was also forbidden to women back then (Yvonne Choquet-Bruhat became the first woman admitted to the Academy in...1979).

Through experiments, she demonstrated Leibniz's theories on kinetic energy. Voltaire, with whom she had a romantic liaison, encouraged her in her research.

Emmy Noether
1882 – 1935

German mathematician

Amalie Emmy Noether was born on March 23, 1882 in Erlangen (Germany).

She had a gift for languages and could have easily taught English or French after passing her exams with flying colors, but she preferred instead to embark on a mathematical course of studies at the University of Erlangen, despite it being largely inaccessible to women: out of one thousand students, only two were women.

Later, having again excelled at her studies, she went on to teach mathematics, despite resistance from her male colleagues, and she did so without status or remuneration, simply because she was a woman. But nothing could stop her, and her classes, which emphasized oral communication and dialogue with students, earned her a special reputation.

In addition to her influence on the mathematicians of the time, she revolutionized the field, and algebra in particular. The theorem in her paper Theory of Ideals in Ring Domains, later called Noether's Theorem, was met with great enthusiasm by researchers.

Albert Einstein called her "The most important creative mathematical genius produced since women gained access to higher education."

Grace Alele-Williams
1932 –

Nigerian Mathematician

Born in 1932 in Niger, **Grace Alele-Williams** studied in Warri and then in Lagos before going to the U.S., where she earned her doctorate in mathematics at the University of Chicago in 1963.

The first Nigerian woman to obtain such a diploma, she then returned to her country, and went on to become head of the University of Benin and the first woman to hold such a high-level, decision-making position.

Her passion for mathematics led her to teaching, which she would continue to do her whole life. As a committed advocate for women's rights, she was particularly dedicated to teaching women.

She has been a member of the African Commission on Mathematics and Vice-President of the Third World Organization for Women in Science.

As part of her struggle on behalf of women's rights, she launched new programs that allow older women working as teachers in elementary schools to receive certification. Whenever she has the chance, she expresses her wish to see many more women in scientific fields and in positions of power.

She is a recipient of the Order of the Niger merit award.

HEDY LAMARR

1914 - 2000
Austrian-American pioneer in transmission methods.
Pioneer Trophy from the *Electronic Frontier Foundation* in 1997,
shared with George Antheil.

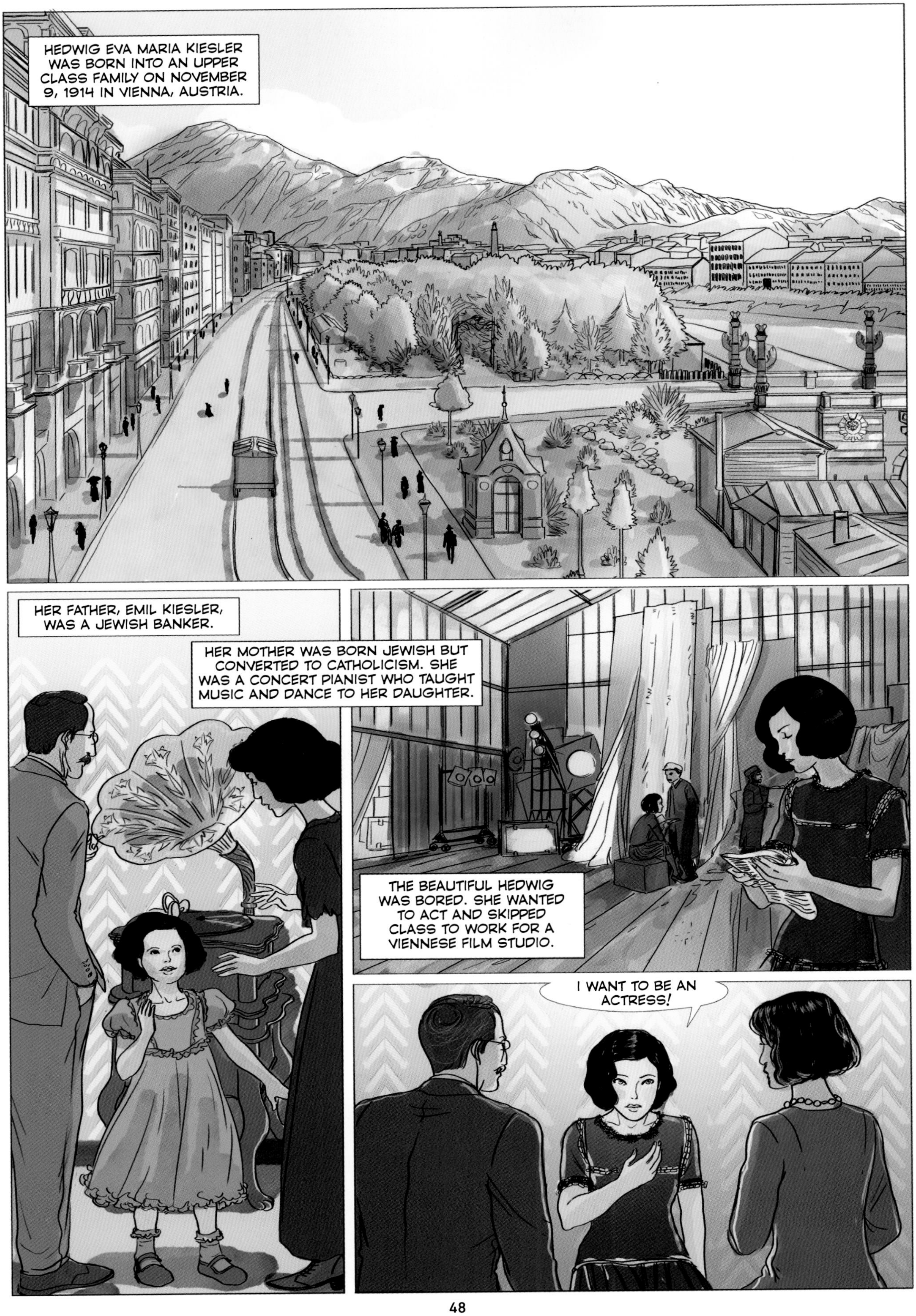
HEDWIG EVA MARIA KIESLER WAS BORN INTO AN UPPER CLASS FAMILY ON NOVEMBER 9, 1914 IN VIENNA, AUSTRIA.
HER FATHER, EMIL KIESLER, WAS A JEWISH BANKER.
HER MOTHER WAS BORN JEWISH BUT CONVERTED TO CATHOLICISM. SHE WAS A CONCERT PIANIST WHO TAUGHT MUSIC AND DANCE TO HER DAUGHTER.
THE BEAUTIFUL HEDWIG WAS BORED. SHE WANTED TO ACT AND SKIPPED CLASS TO WORK FOR A VIENNESE FILM STUDIO.
I WANT TO BE AN ACTRESS!

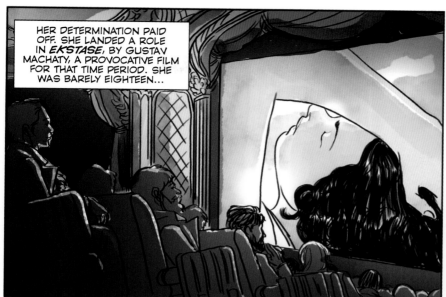

HER DETERMINATION PAID OFF. SHE LANDED A ROLE IN *EKSTASE*, BY GUSTAV MACHATY, A PROVOCATIVE FILM FOR THAT TIME PERIOD. SHE WAS BARELY EIGHTEEN...

SCANDALOUS, HEDY!

DESPITE THE SCANDAL, HEDY KEPT ACTING, ALSO APPEARING ON THE VIENNA STAGE IN *SISSI*.

MORE FLOWERS, MISS!

WE'RE OUT OF VASES! JUST PUT THEM IN THE SINK.

MISS HEDY! OVER HERE!

MISS KIESLER, YOU WERE MARVELOUS! PLEASE DO ME THE HONOR OF DINING WITH ME.

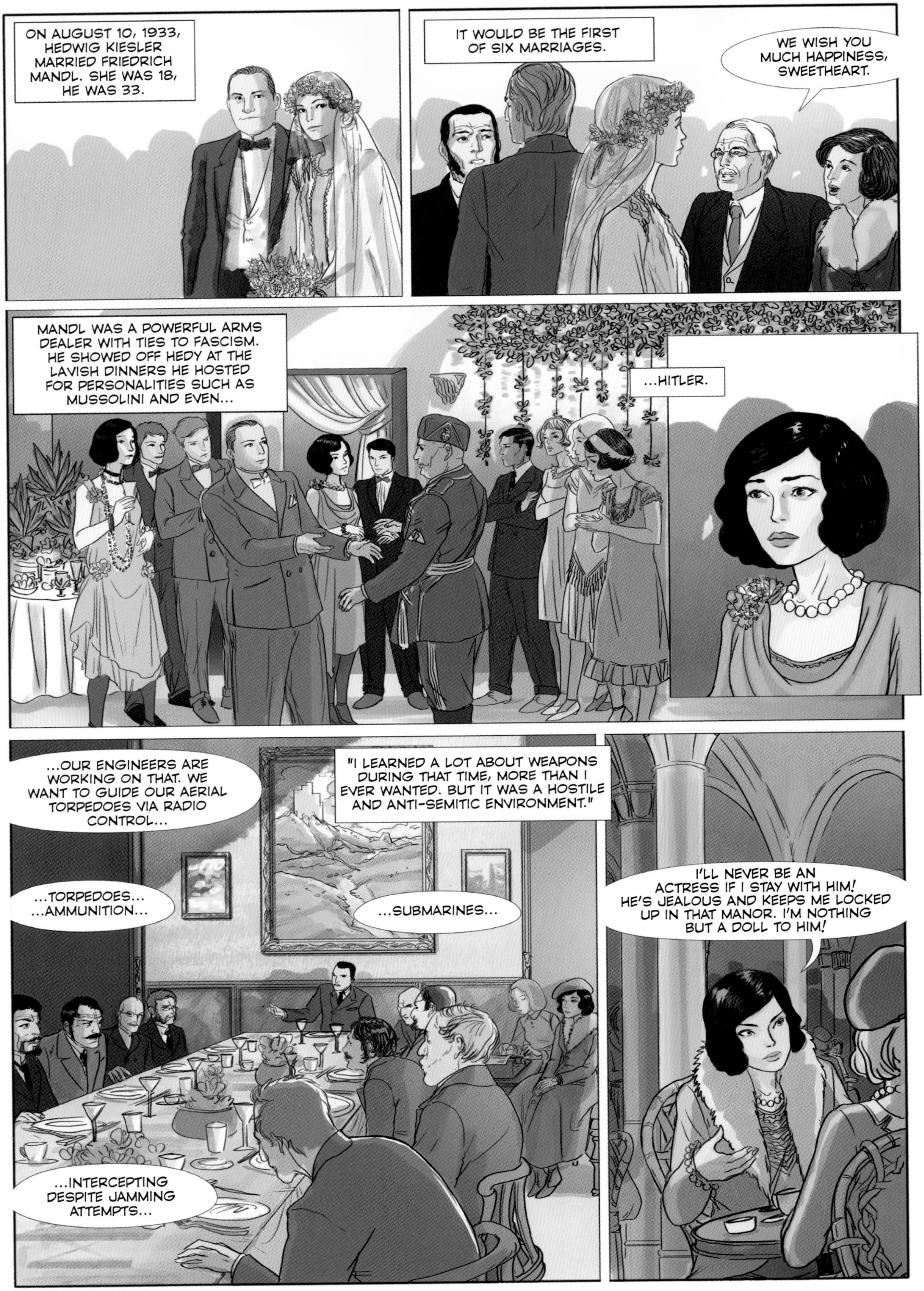
ON AUGUST 10, 1933, HEDWIG KIESLER MARRIED FRIEDRICH MANDL. SHE WAS 18, HE WAS 33.
IT WOULD BE THE FIRST OF SIX MARRIAGES.
WE WISH YOU MUCH HAPPINESS, SWEETHEART.
MANDL WAS A POWERFUL ARMS DEALER WITH TIES TO FASCISM. HE SHOWED OFF HEDY AT THE LAVISH DINNERS HE HOSTED FOR PERSONALITIES SUCH AS MUSSOLINI AND EVEN...
...HITLER.
...OUR ENGINEERS ARE WORKING ON THAT. WE WANT TO GUIDE OUR AERIAL TORPEDOES VIA RADIO CONTROL...
"I LEARNED A LOT ABOUT WEAPONS DURING THAT TIME, MORE THAN I EVER WANTED. BUT IT WAS A HOSTILE AND ANTI-SEMITIC ENVIRONMENT."
...TORPEDOES... ...AMMUNITION...
...SUBMARINES...
...INTERCEPTING DESPITE JAMMING ATTEMPTS...
I'LL NEVER BE AN ACTRESS IF I STAY WITH HIM! HE'S JEALOUS AND KEEPS ME LOCKED UP IN THAT MANOR. I'M NOTHING BUT A DOLL TO HIM!

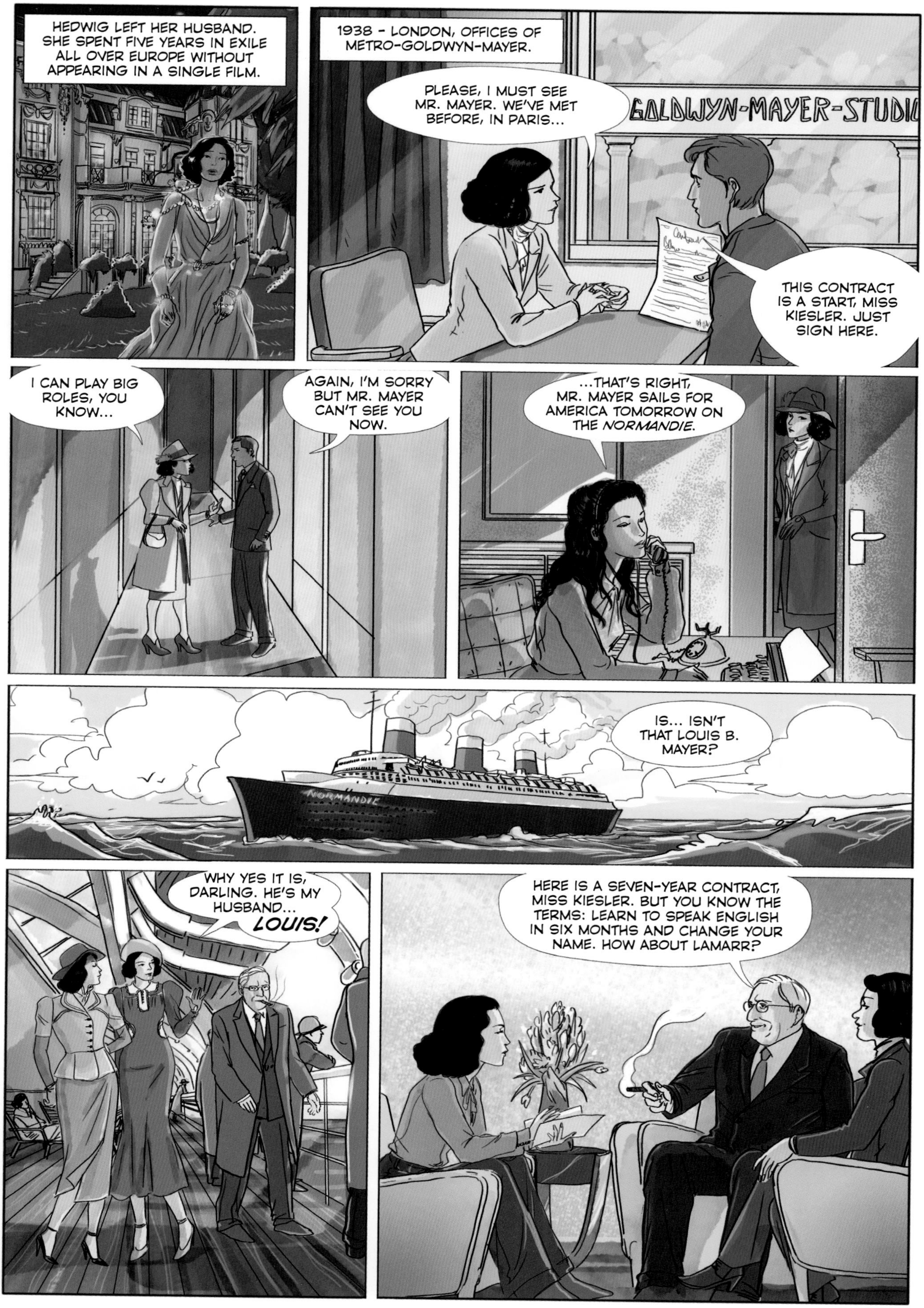
HEDWIG LEFT HER HUSBAND. SHE SPENT FIVE YEARS IN EXILE ALL OVER EUROPE WITHOUT APPEARING IN A SINGLE FILM.
1938 - LONDON, OFFICES OF METRO-GOLDWYN-MAYER.
PLEASE, I MUST SEE MR. MAYER. WE'VE MET BEFORE, IN PARIS...
GOLDWYN-MAYER-STUDIO
THIS CONTRACT IS A START, MISS KIESLER. JUST SIGN HERE.
I CAN PLAY BIG ROLES, YOU KNOW...
AGAIN, I'M SORRY BUT MR. MAYER CAN'T SEE YOU NOW.
...THAT'S RIGHT, MR. MAYER SAILS FOR AMERICA TOMORROW ON THE *NORMANDIE*.
IS... ISN'T THAT LOUIS B. MAYER?
NORMANDIE
WHY YES IT IS, DARLING. HE'S MY HUSBAND... **LOUIS!**
HERE IS A SEVEN-YEAR CONTRACT, MISS KIESLER. BUT YOU KNOW THE TERMS: LEARN TO SPEAK ENGLISH IN SIX MONTHS AND CHANGE YOUR NAME. HOW ABOUT LAMARR?

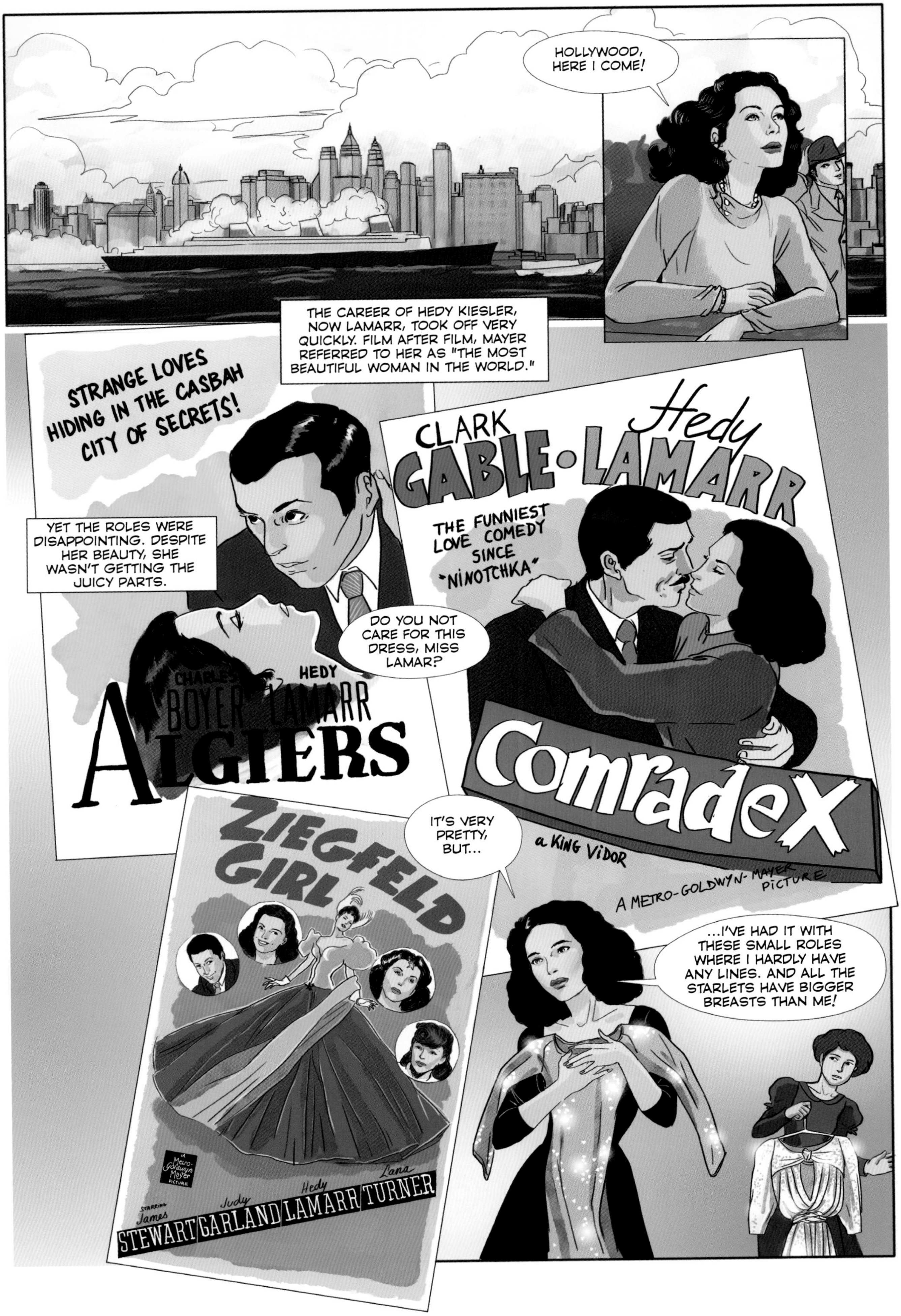
HOLLYWOOD, HERE I COME!
THE CAREER OF HEDY KIESLER, NOW LAMARR, TOOK OFF VERY QUICKLY. FILM AFTER FILM, MAYER REFERRED TO HER AS "THE MOST BEAUTIFUL WOMAN IN THE WORLD."
STRANGE LOVES HIDING IN THE CASBAH CITY OF SECRETS!
CLARK GABLE · Hedy LAMARR
THE FUNNIEST LOVE COMEDY SINCE "NINOTCHKA"
YET THE ROLES WERE DISAPPOINTING. DESPITE HER BEAUTY, SHE WASN'T GETTING THE JUICY PARTS.
DO YOU NOT CARE FOR THIS DRESS, MISS LAMAR?
CHARLES BOYER HEDY LAMARR
ALGIERS
Comradex
a KING VIDOR
A METRO-GOLDWYN-MAYER PICTURE
ZIEGFELD GIRL
IT'S VERY PRETTY, BUT...
...I'VE HAD IT WITH THESE SMALL ROLES WHERE I HARDLY HAVE ANY LINES. AND ALL THE STARLETS HAVE BIGGER BREASTS THAN ME!
STARRING James STEWART Judy GARLAND Hedy LAMARR Lana TURNER

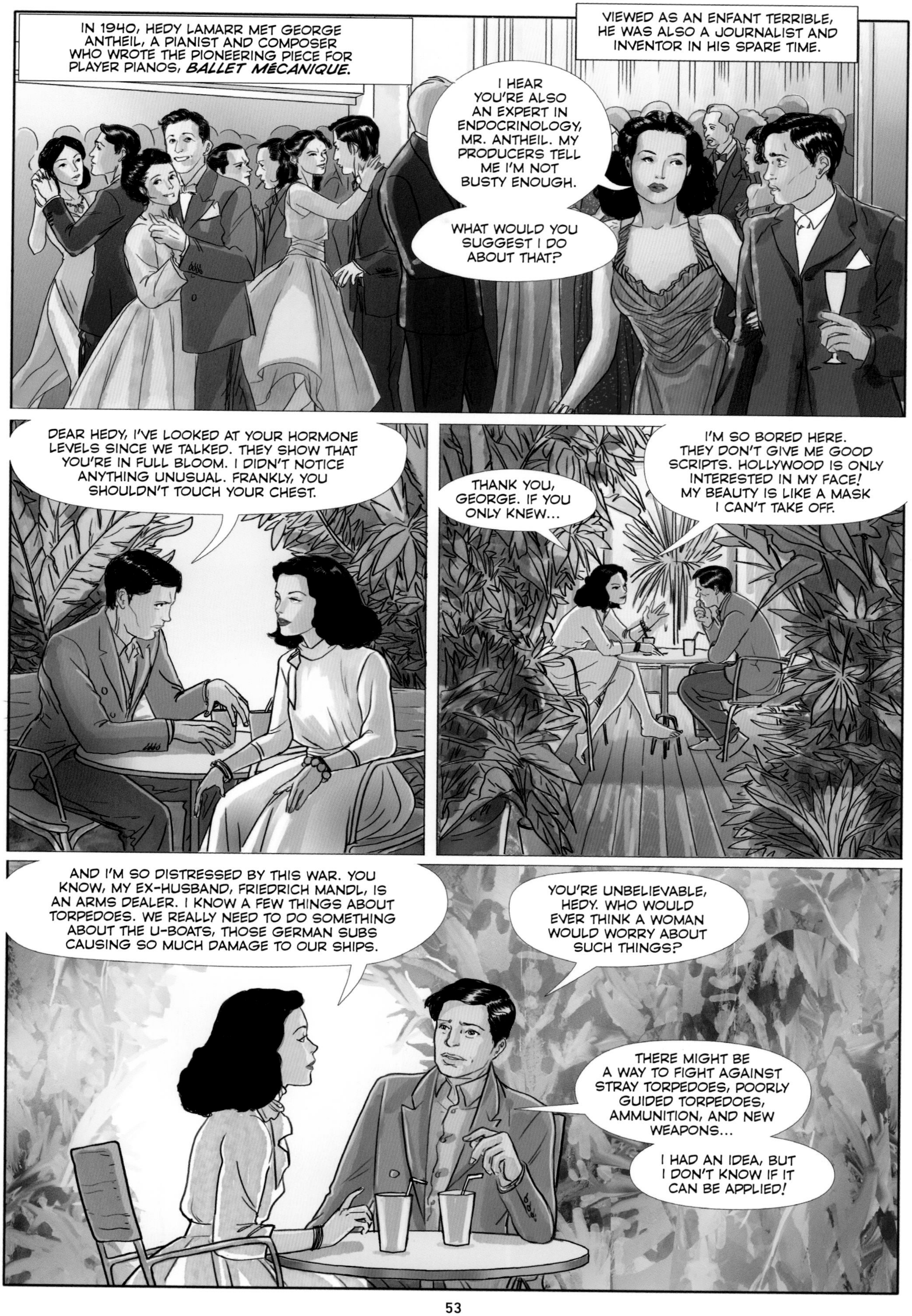
IN 1940, HEDY LAMARR MET GEORGE ANTHEIL, A PIANIST AND COMPOSER WHO WROTE THE PIONEERING PIECE FOR PLAYER PIANOS, *BALLET MÉCANIQUE*.
VIEWED AS AN ENFANT TERRIBLE, HE WAS ALSO A JOURNALIST AND INVENTOR IN HIS SPARE TIME.
I HEAR YOU'RE ALSO AN EXPERT IN ENDOCRINOLOGY, MR. ANTHEIL. MY PRODUCERS TELL ME I'M NOT BUSTY ENOUGH.
WHAT WOULD YOU SUGGEST I DO ABOUT THAT?
DEAR HEDY, I'VE LOOKED AT YOUR HORMONE LEVELS SINCE WE TALKED. THEY SHOW THAT YOU'RE IN FULL BLOOM. I DIDN'T NOTICE ANYTHING UNUSUAL. FRANKLY, YOU SHOULDN'T TOUCH YOUR CHEST.
THANK YOU, GEORGE. IF YOU ONLY KNEW...
I'M SO BORED HERE. THEY DON'T GIVE ME GOOD SCRIPTS. HOLLYWOOD IS ONLY INTERESTED IN MY FACE! MY BEAUTY IS LIKE A MASK I CAN'T TAKE OFF.
AND I'M SO DISTRESSED BY THIS WAR. YOU KNOW, MY EX-HUSBAND, FRIEDRICH MANDL, IS AN ARMS DEALER. I KNOW A FEW THINGS ABOUT TORPEDOES. WE REALLY NEED TO DO SOMETHING ABOUT THE U-BOATS, THOSE GERMAN SUBS CAUSING SO MUCH DAMAGE TO OUR SHIPS.
YOU'RE UNBELIEVABLE, HEDY. WHO WOULD EVER THINK A WOMAN WOULD WORRY ABOUT SUCH THINGS?
THERE MIGHT BE A WAY TO FIGHT AGAINST STRAY TORPEDOES, POORLY GUIDED TORPEDOES, AMMUNITION, AND NEW WEAPONS...
I HAD AN IDEA, BUT I DON'T KNOW IF IT CAN BE APPLIED!

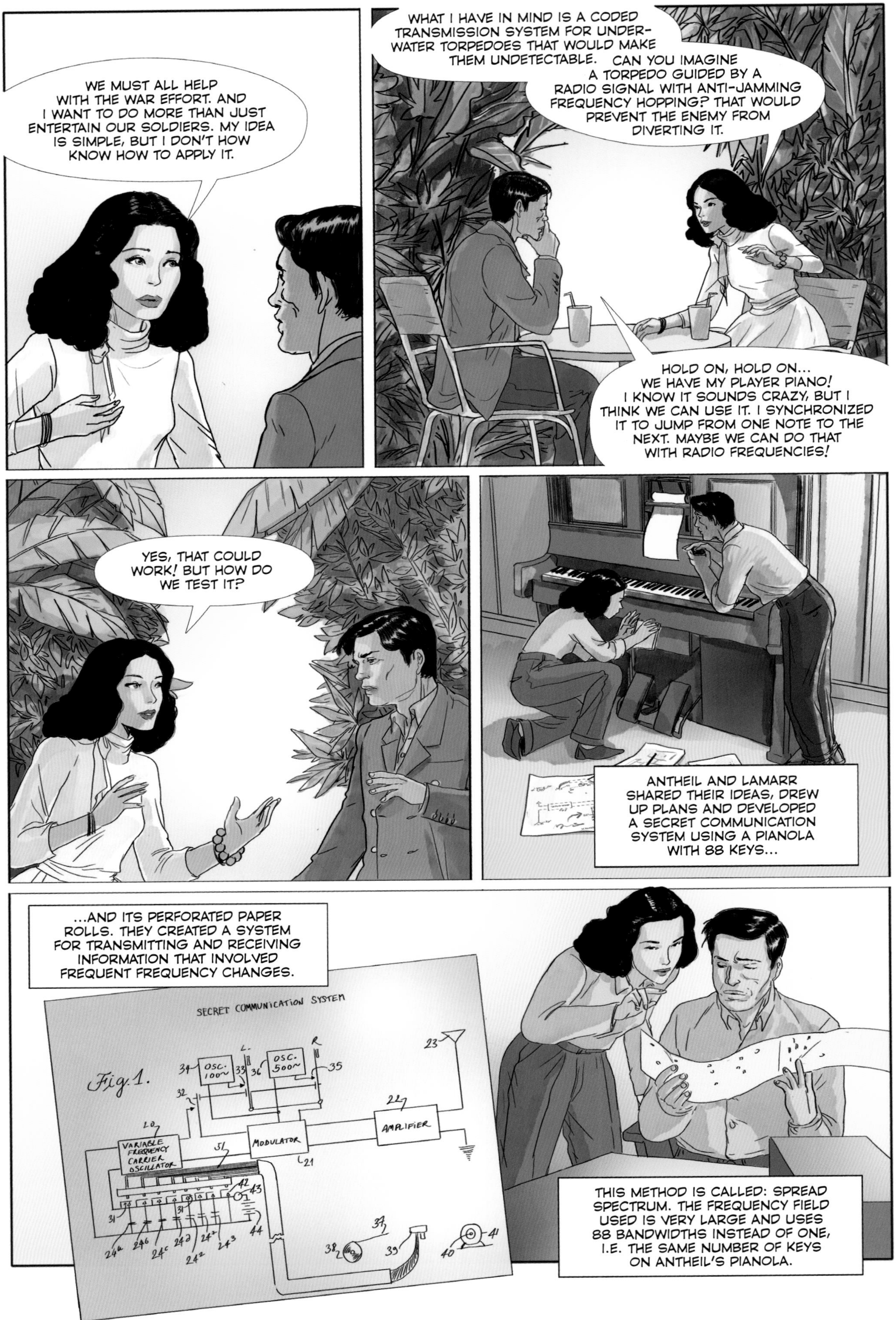
WE MUST ALL HELP WITH THE WAR EFFORT. AND I WANT TO DO MORE THAN JUST ENTERTAIN OUR SOLDIERS. MY IDEA IS SIMPLE, BUT I DON'T HOW KNOW HOW TO APPLY IT.
WHAT I HAVE IN MIND IS A CODED TRANSMISSION SYSTEM FOR UNDER-WATER TORPEDOES THAT WOULD MAKE THEM UNDETECTABLE. CAN YOU IMAGINE A TORPEDO GUIDED BY A RADIO SIGNAL WITH ANTI-JAMMING FREQUENCY HOPPING? THAT WOULD PREVENT THE ENEMY FROM DIVERTING IT.
HOLD ON, HOLD ON... WE HAVE MY PLAYER PIANO! I KNOW IT SOUNDS CRAZY, BUT I THINK WE CAN USE IT. I SYNCHRONIZED IT TO JUMP FROM ONE NOTE TO THE NEXT. MAYBE WE CAN DO THAT WITH RADIO FREQUENCIES!
YES, THAT COULD WORK! BUT HOW DO WE TEST IT?
ANTHEIL AND LAMARR SHARED THEIR IDEAS, DREW UP PLANS AND DEVELOPED A SECRET COMMUNICATION SYSTEM USING A PIANOLA WITH 88 KEYS...
...AND ITS PERFORATED PAPER ROLLS. THEY CREATED A SYSTEM FOR TRANSMITTING AND RECEIVING INFORMATION THAT INVOLVED FREQUENT FREQUENCY CHANGES.
SECRET COMMUNICATION SYSTEM
Fig. 1.
OSC. 100~
OSC. 500~
AMPLIFIER
MODULATOR
VARIABLE FREQUENCY CARRIER OSCILLATOR
THIS METHOD IS CALLED: SPREAD SPECTRUM. THE FREQUENCY FIELD USED IS VERY LARGE AND USES 88 BANDWIDTHS INSTEAD OF ONE, I.E. THE SAME NUMBER OF KEYS ON ANTHEIL'S PIANOLA.

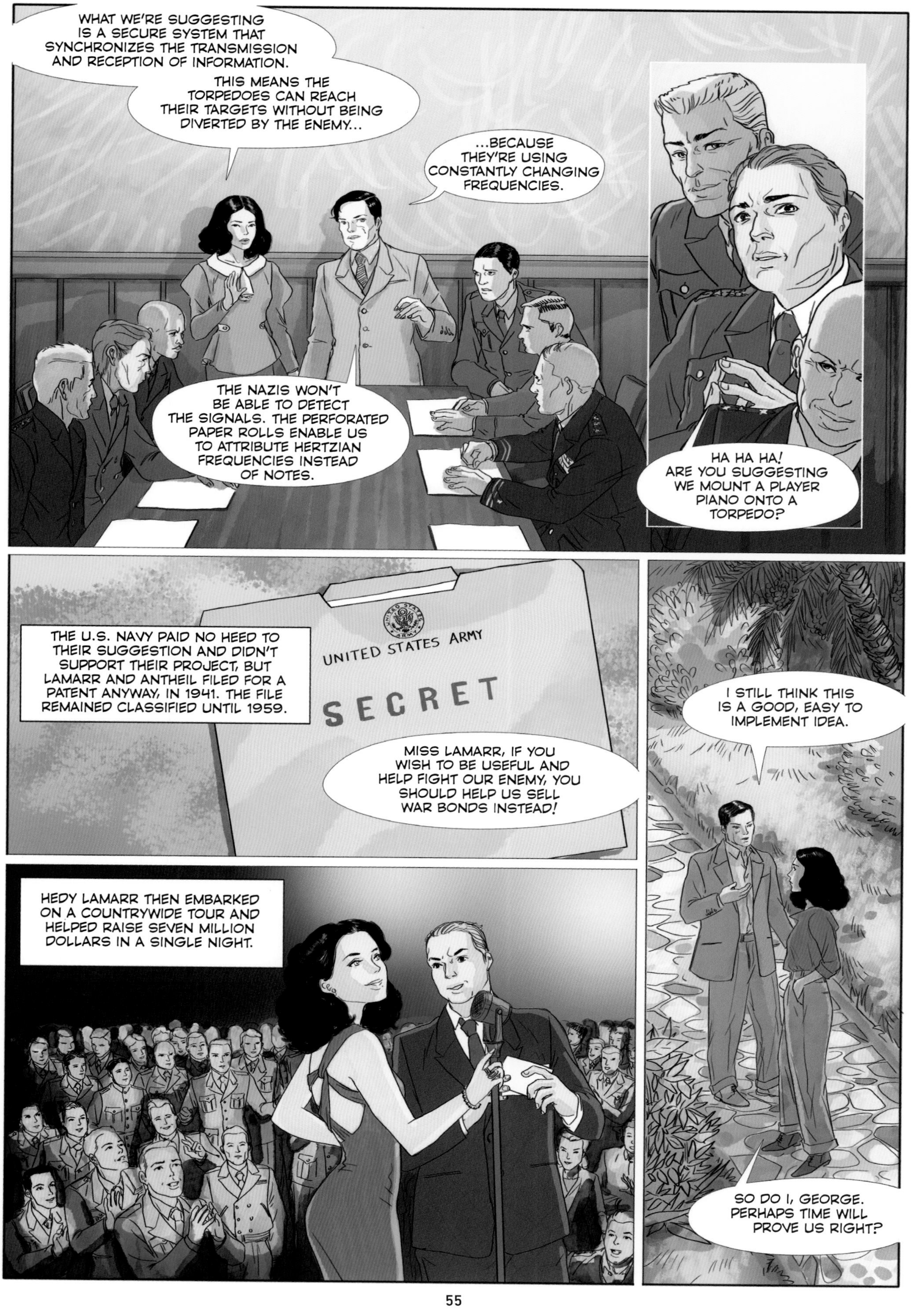
WHAT WE'RE SUGGESTING IS A SECURE SYSTEM THAT SYNCHRONIZES THE TRANSMISSION AND RECEPTION OF INFORMATION.
THIS MEANS THE TORPEDOES CAN REACH THEIR TARGETS WITHOUT BEING DIVERTED BY THE ENEMY...
...BECAUSE THEY'RE USING CONSTANTLY CHANGING FREQUENCIES.
THE NAZIS WON'T BE ABLE TO DETECT THE SIGNALS. THE PERFORATED PAPER ROLLS ENABLE US TO ATTRIBUTE HERTZIAN FREQUENCIES INSTEAD OF NOTES.
HA HA HA! ARE YOU SUGGESTING WE MOUNT A PLAYER PIANO ONTO A TORPEDO?
THE U.S. NAVY PAID NO HEED TO THEIR SUGGESTION AND DIDN'T SUPPORT THEIR PROJECT, BUT LAMARR AND ANTHEIL FILED FOR A PATENT ANYWAY, IN 1941. THE FILE REMAINED CLASSIFIED UNTIL 1959.
UNITED STATES ARMY
SECRET
MISS LAMARR, IF YOU WISH TO BE USEFUL AND HELP FIGHT OUR ENEMY, YOU SHOULD HELP US SELL WAR BONDS INSTEAD!
I STILL THINK THIS IS A GOOD, EASY TO IMPLEMENT IDEA.
HEDY LAMARR THEN EMBARKED ON A COUNTRYWIDE TOUR AND HELPED RAISE SEVEN MILLION DOLLARS IN A SINGLE NIGHT.
SO DO I, GEORGE. PERHAPS TIME WILL PROVE US RIGHT?

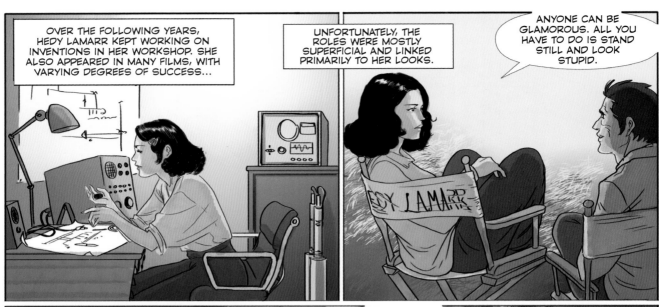

OVER THE FOLLOWING YEARS, HEDY LAMARR KEPT WORKING ON INVENTIONS IN HER WORKSHOP. SHE ALSO APPEARED IN MANY FILMS, WITH VARYING DEGREES OF SUCCESS...

UNFORTUNATELY, THE ROLES WERE MOSTLY SUPERFICIAL AND LINKED PRIMARILY TO HER LOOKS.

ANYONE CAN BE GLAMOROUS. ALL YOU HAVE TO DO IS STAND STILL AND LOOK STUPID.

SHE WAS OVERLY WARY AND MADE THE MISTAKE OF TURNING DOWN TWO FILMS THAT WENT ON TO BE BIG HITS: *GASLIGHT* AND *CASABLANCA*.

DESPITE THE SUCCESS OF CECIL B. DEMILLE'S *SAMSON AND DALILAH* IN 1959, THE ACTRESS NEVER REACHED THE FAME SHE'D HOPED FOR. HER CAREER BEGAN TO DECLINE AND SHE QUIT ACTING IN 1958.

I WAS THE BIGGEST STAR IN HOLLYWOOD, BUT I WAS "DIFFICULT."

"THE LADDER OF SUCCESS IN HOLLYWOOD IS USUALLY A PRESS AGENT, ACTOR, DIRECTOR, PRODUCER, LEADING MAN; AND YOU ARE A STAR IF YOU SLEEP WITH EACH OF THEM IN THAT ORDER. CRUDE, BUT TRUE."

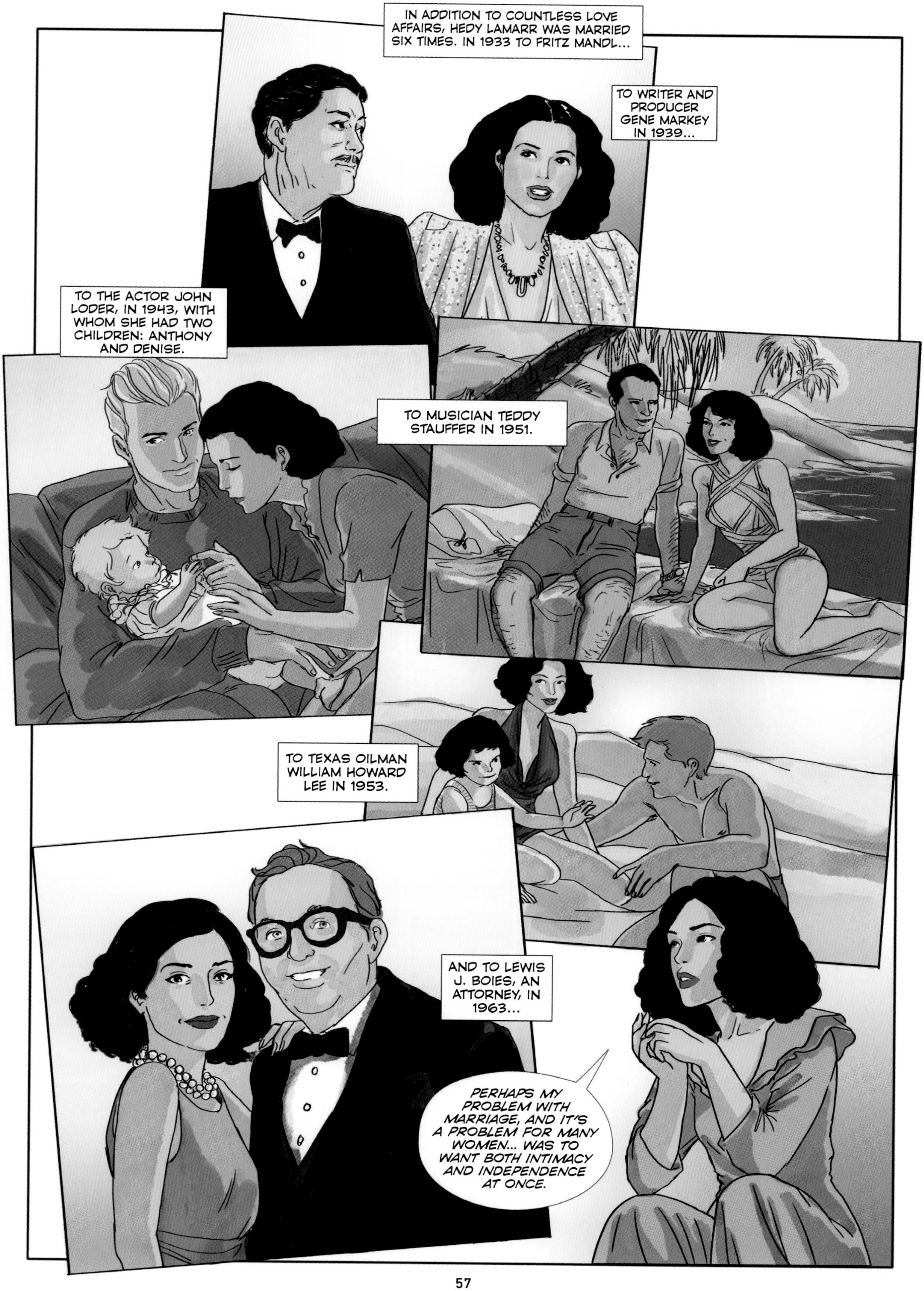
IN ADDITION TO COUNTLESS LOVE AFFAIRS, HEDY LAMARR WAS MARRIED SIX TIMES. IN 1933 TO FRITZ MANDL...
TO WRITER AND PRODUCER GENE MARKEY IN 1939...
TO THE ACTOR JOHN LODER, IN 1943, WITH WHOM SHE HAD TWO CHILDREN: ANTHONY AND DENISE.
TO MUSICIAN TEDDY STAUFFER IN 1951.
TO TEXAS OILMAN WILLIAM HOWARD LEE IN 1953.
AND TO LEWIS J. BOIES, AN ATTORNEY, IN 1963...
PERHAPS MY PROBLEM WITH MARRIAGE, AND IT'S A PROBLEM FOR MANY WOMEN... WAS TO WANT BOTH INTIMACY AND INDEPENDENCE AT ONCE.

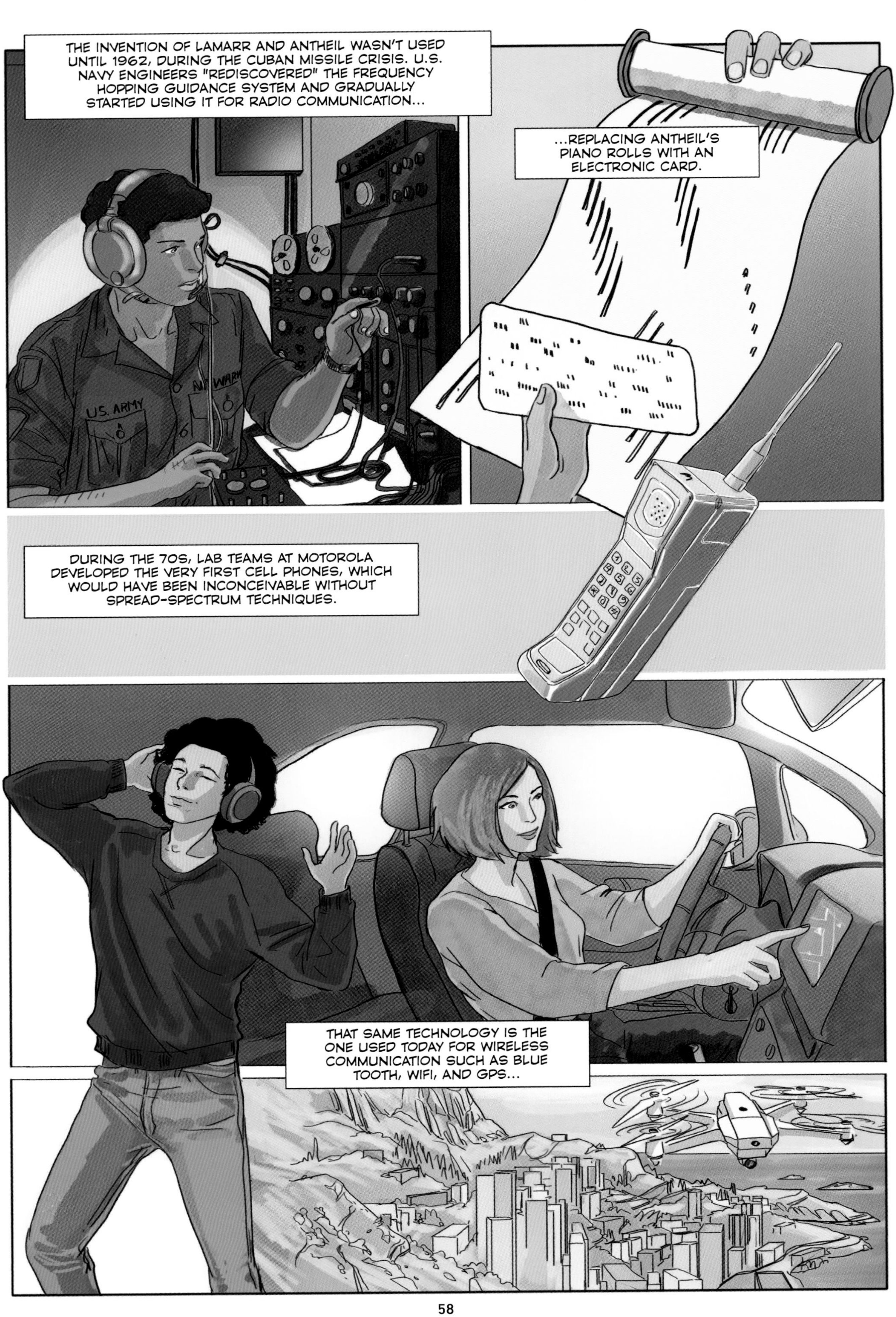
THE INVENTION OF LAMARR AND ANTHEIL WASN'T USED UNTIL 1962, DURING THE CUBAN MISSILE CRISIS. U.S. NAVY ENGINEERS "REDISCOVERED" THE FREQUENCY HOPPING GUIDANCE SYSTEM AND GRADUALLY STARTED USING IT FOR RADIO COMMUNICATION...
U.S. ARMY
...REPLACING ANTHEIL'S PIANO ROLLS WITH AN ELECTRONIC CARD.
DURING THE 70S, LAB TEAMS AT MOTOROLA DEVELOPED THE VERY FIRST CELL PHONES, WHICH WOULD HAVE BEEN INCONCEIVABLE WITHOUT SPREAD-SPECTRUM TECHNIQUES.
THAT SAME TECHNOLOGY IS THE ONE USED TODAY FOR WIRELESS COMMUNICATION SUCH AS BLUE TOOTH, WIFI, AND GPS...

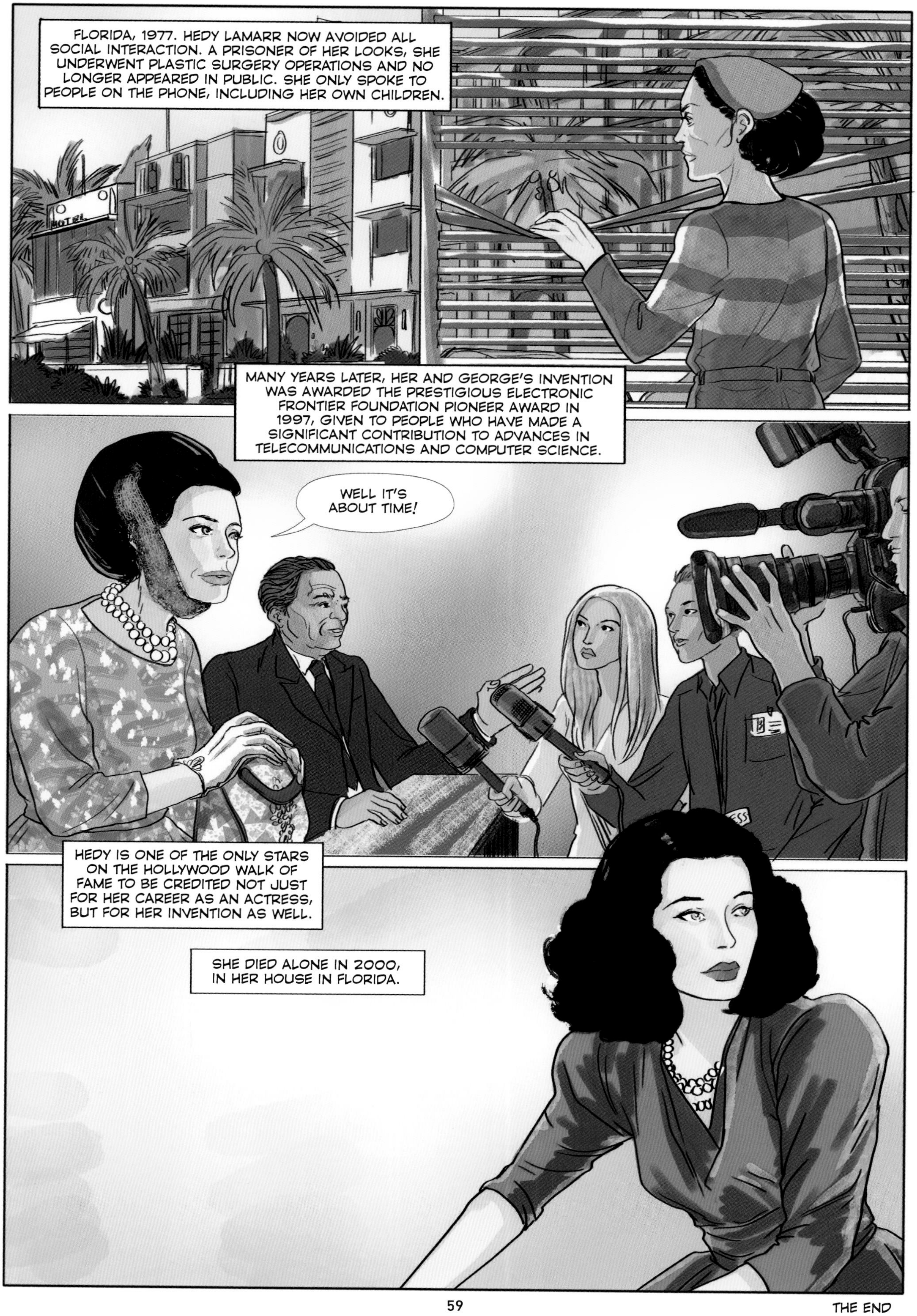
FLORIDA, 1977. HEDY LAMARR NOW AVOIDED ALL SOCIAL INTERACTION. A PRISONER OF HER LOOKS, SHE UNDERWENT PLASTIC SURGERY OPERATIONS AND NO LONGER APPEARED IN PUBLIC. SHE ONLY SPOKE TO PEOPLE ON THE PHONE, INCLUDING HER OWN CHILDREN.
HOTEL
MANY YEARS LATER, HER AND GEORGE'S INVENTION WAS AWARDED THE PRESTIGIOUS ELECTRONIC FRONTIER FOUNDATION PIONEER AWARD IN 1997, GIVEN TO PEOPLE WHO HAVE MADE A SIGNIFICANT CONTRIBUTION TO ADVANCES IN TELECOMMUNICATIONS AND COMPUTER SCIENCE.
WELL IT'S ABOUT TIME!
HEDY IS ONE OF THE ONLY STARS ON THE HOLLYWOOD WALK OF FAME TO BE CREDITED NOT JUST FOR HER CAREER AS AN ACTRESS, BUT FOR HER INVENTION AS WELL.
SHE DIED ALONE IN 2000, IN HER HOUSE IN FLORIDA.

THE END

Katherine Coleman Goble Johnson
1918 – 2020

American mathematician and astrophysicist

Born in 1918 in West Virginia, to a farmer father and a school teacher mother, **Katherine Coleman** demonstrated an inclination for studies at an early age.

As the education system in her county did not offer classes to African-American students past the 8th grade, young Katherine had to attend high school at West Virginia State College in the town of Institute, from which she graduated at the age of fourteen.

Her gift for mathematics landed her a teaching job shortly after graduation, but she quit to attend graduate school, hoping to embark on a career as a research mathematician. In 1952, she responded to an announcement put out by NACA (later NASA), which was seeking to integrate African-American women.

There, she became one of the brains of the mathematical calculations department, enduring racism and sexism while asserting herself as the only female member of a team working on flights, trajectories and aerodynamics.

Among many prestigious assignments, she calculated the trajectory of the Apollo 11 mission to the Moon. In 2015, the little Black girl who became a NASA engineer was awarded the Presidential Medal of Freedom, the highest civilian decoration in the United States.

Marthe Gautier
1925 –

French pediatrician and researcher

Born in 1925 into a family of farmers, **Marthe Gautier** decided early on to study medicine, following in the footsteps of her sister, a pediatrician who died at a young age during WWII fighting.

Initially a heart disease researcher, she excelled in cell culture, which she developed at the Trousseau hospital in Paris with very few resources, a lot of inventiveness, and her own personal techniques.

She soon discovered that the cells of normal children had 46 chromosomes, whereas the tissues of children with Down Syndrome had one extra chromosome. One of her colleagues, Jérome Lejeune, a researcher in the same department at the CNRS, France's leading research Institute, then took an interest in her work and allowed her to work in better conditions, with more effective and recent equipment. Unfortunately, he took it upon himself to announce the discovery to the scientific world during a genetics seminar in Montreal, unfairly leaving Marthe Gautier in the shadows.

Not until 2014 did the ethics committee of INSERM (the French National Institute of Health and Medical Research) declare: "The discovery of Down Syndrome could not have been made without the essential contributions of Raymond Turpin and Marthe Gautier; it is regrettable that both their names were not automatically associated with this discovery, both in the announcement and in the attribution of various honors."

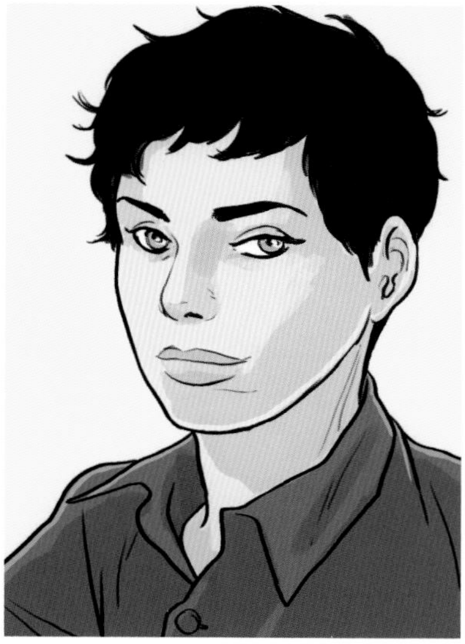

Maryam Mirzakhani
1977 – 2017

Iranian mathematician
Fields Medal 2014

Born in 1977 in Iran, **Maryam Mirzakhani** earned a Bachelors degree in mathematics in Tehran, in an academic system that was increasingly open to women, and then moved to the U.S., where she earned a doctorate in mathematics at Harvard in 2004.

She became the first and only woman to win the Fields Medal, know as the Nobel Prize in Mathematics and awarded every four years, for her work in geometry and the calculation of hyperbolic surfaces--such as the surface of a donut, for example.

Maryam had a very particular way of approaching and teaching mathematics, which she viewed a bit like an artistic subject. Where most people saw formulas and symbols, she would see scenes; she liked to make doodles of natural elements such as flowers, and she enjoyed solving puzzles with an ease, optimism and joy that she passed on to her students and colleagues, allowing them to push past their own limits.

Sadly, she died prematurely of breast cancer at the age of 40, leaving the world of mathematics in mourning.

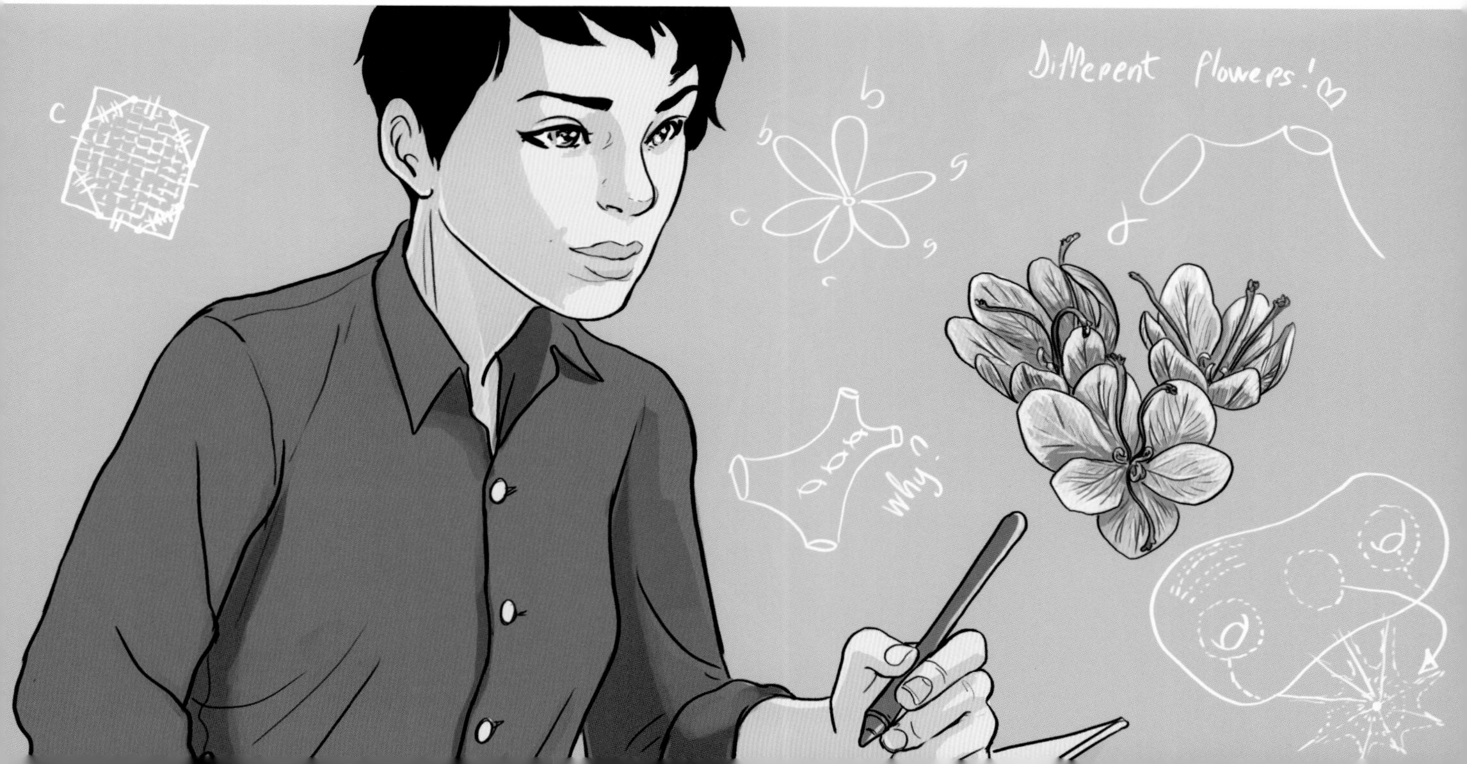
Different flowers!
why?

ROSALIND FRANKLIN

1920 -1958
English pioneer in the field of molecular biology and DNA research.
Awarded the Louisa Gross Horwitz prize posthumously in 2008.

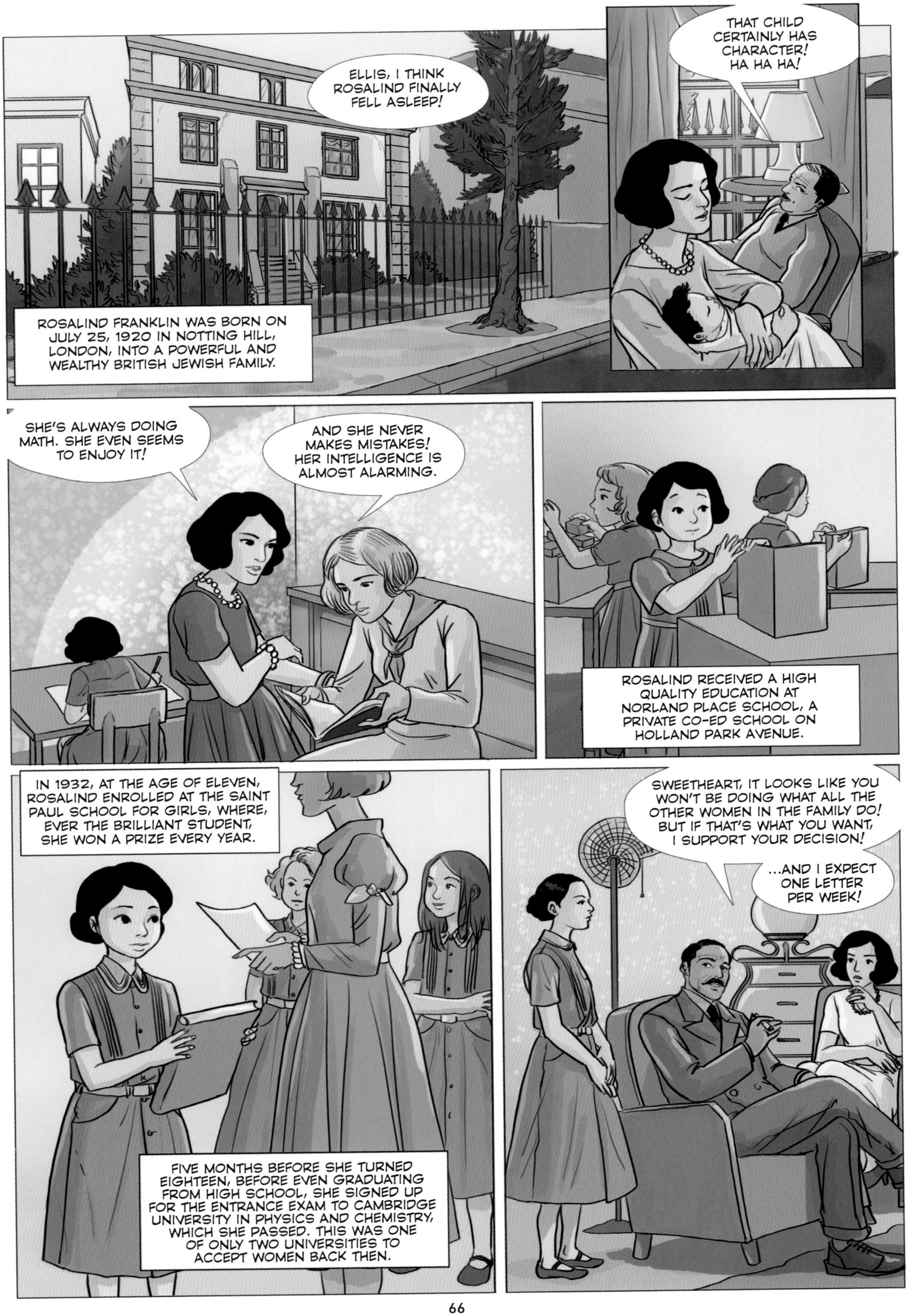
ELLIS, I THINK ROSALIND FINALLY FELL ASLEEP!
ROSALIND FRANKLIN WAS BORN ON JULY 25, 1920 IN NOTTING HILL, LONDON, INTO A POWERFUL AND WEALTHY BRITISH JEWISH FAMILY.
THAT CHILD CERTAINLY HAS CHARACTER! HA HA HA!
SHE'S ALWAYS DOING MATH. SHE EVEN SEEMS TO ENJOY IT!
AND SHE NEVER MAKES MISTAKES! HER INTELLIGENCE IS ALMOST ALARMING.
ROSALIND RECEIVED A HIGH QUALITY EDUCATION AT NORLAND PLACE SCHOOL, A PRIVATE CO-ED SCHOOL ON HOLLAND PARK AVENUE.
IN 1932, AT THE AGE OF ELEVEN, ROSALIND ENROLLED AT THE SAINT PAUL SCHOOL FOR GIRLS, WHERE, EVER THE BRILLIANT STUDENT, SHE WON A PRIZE EVERY YEAR.
FIVE MONTHS BEFORE SHE TURNED EIGHTEEN, BEFORE EVEN GRADUATING FROM HIGH SCHOOL, SHE SIGNED UP FOR THE ENTRANCE EXAM TO CAMBRIDGE UNIVERSITY IN PHYSICS AND CHEMISTRY, WHICH SHE PASSED. THIS WAS ONE OF ONLY TWO UNIVERSITIES TO ACCEPT WOMEN BACK THEN.
SWEETHEART, IT LOOKS LIKE YOU WON'T BE DOING WHAT ALL THE OTHER WOMEN IN THE FAMILY DO! BUT IF THAT'S WHAT YOU WANT, I SUPPORT YOUR DECISION!
...AND I EXPECT ONE LETTER PER WEEK!

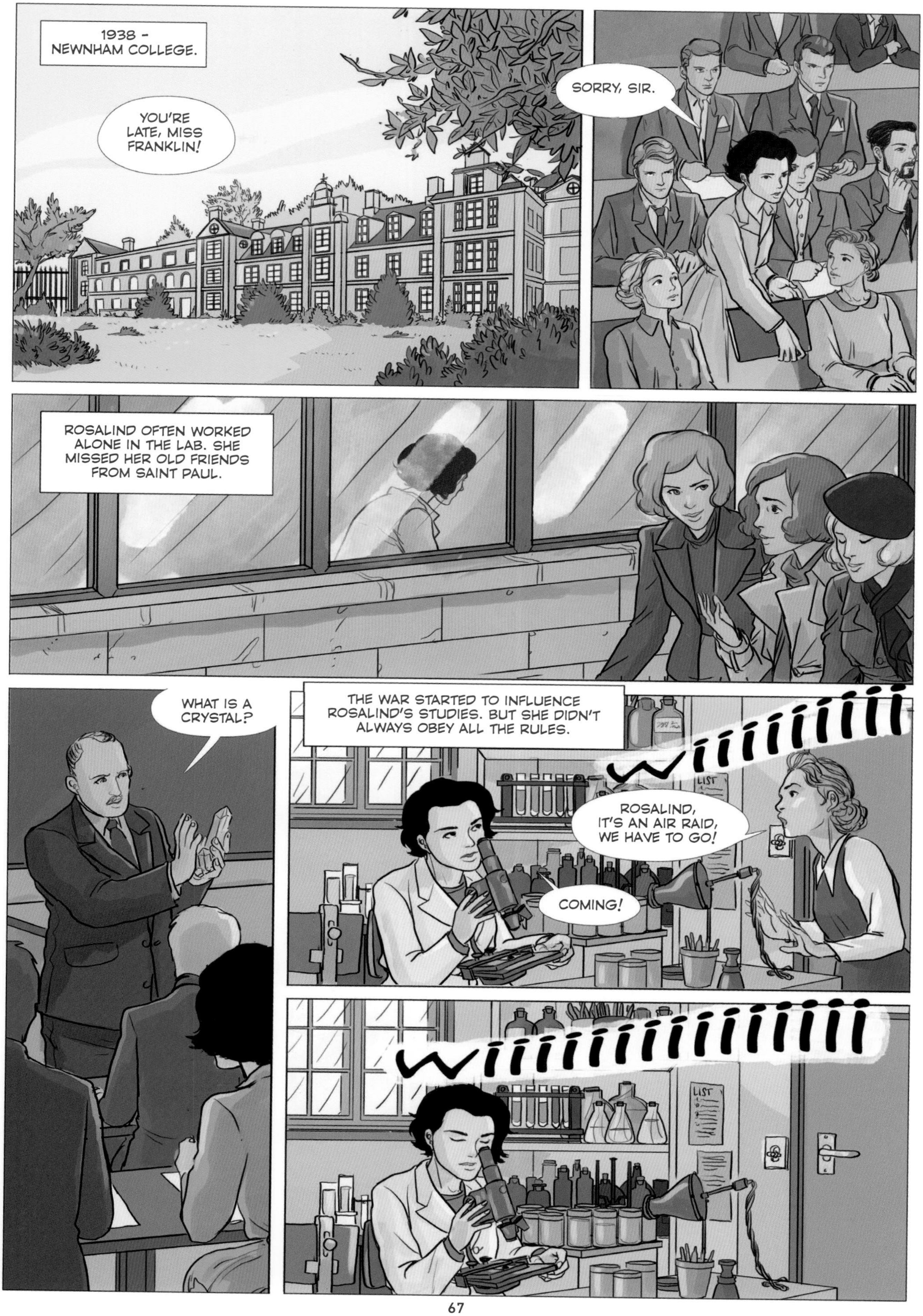
1938 - NEWNHAM COLLEGE.
YOU'RE LATE, MISS FRANKLIN!
SORRY, SIR.
ROSALIND OFTEN WORKED ALONE IN THE LAB. SHE MISSED HER OLD FRIENDS FROM SAINT PAUL.
WHAT IS A CRYSTAL?
THE WAR STARTED TO INFLUENCE ROSALIND'S STUDIES. BUT SHE DIDN'T ALWAYS OBEY ALL THE RULES.
WIIIIIIIII
ROSALIND, IT'S AN AIR RAID, WE HAVE TO GO!
COMING!
WIIIIIIIIIIIIIII

1940: THE THIRD YEAR WAS THE ONE EMPHASIZING PHYSICS AND CHEMISTRY, BUT FOR ROSALIND, IT WAS THE YEAR OF THE LONDON BLITZ.

"DEAR PARENTS, VIRTUALLY THE ENTIRE LAB IS GONE. BIOCHEMISTRY IS NOW HANDLED BY PROFESSORS OF GERMAN ANCESTRY THAT CAN'T JOIN THE ARMY..."

"...CAMBRIDGE IS TAKING IN WAR REFUGEES. I HAD THE GOOD FORTUNE OF MEETING ADRIENNE WEILL, A FRENCH SCIENTIST, WHO GAVE A FASCINATING LECTURE, IN FRENCH, ON MARIE CURIE."

"...SHE WAS BOTH HER STUDENT AND CO-RESEARCHER. I'M SO IMPRESSED!"

I'M ANXIOUS ABOUT THE EXAMS. I'M NOT GOOD ENOUGH...

OF COURSE YOU ARE, ROSALIND! THE WORK YOU'VE DONE ON COAL IS BRILLIANT.

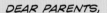

DEAR PARENTS,

IN THE END, I DID GET MY DIPLOMA, CUM LAUDE.

I COULD HAVE DONE BETTER BUT I HAVEN'T BEEN SLEEPING WELL, AND I HAVE A COLD THAT PREVENTED ME FROM PERFORMING AT MY HIGHEST LEVEL.

I CAN'T WAIT TO SEE YOU FOR THE HOLIDAYS.

MUCH LOVE,

ROSALIND

"DEAR PARENTS, I RECEIVED A FELLOWSHIP GRANT AND AM GOING TO STAY ONE MORE YEAR, WORKING UNDER PROFESSOR DAINTON. I AM NOW ONE OF THE VOLUNTEER FIREFIGHTERS AT CAMBRIDGE. SINCE SO MANY MEN ARE IN THE WAR, WE WOMEN HAVE TO TAKE THEIR PLACE... EVEN THOUGH PEOPLE DON'T SEEM TO APPRECIATE IT MUCH..."

...THE TIMES EVEN MADE A FUSS ABOUT THAT! THEY RELEASED AN ARTICLE THAT SAID THAT, "THE BARRIERS ERECTED AGAINST THE WORK OF WOMEN ARE UNACCEPTABLE AND IRRATIONAL! WOMEN SHOULD NOT HAVE TO TAKE A STEP BACKWARDS AFTER THIS!"

ROSALIND, YOU SHOULD BE PROUD OF YOURSELF. THIS GRANT WILL ENABLE YOU TO STUDY AT CAMBRIDGE UNIVERSITY UNDER PROFESSOR RONALD NORRISH.

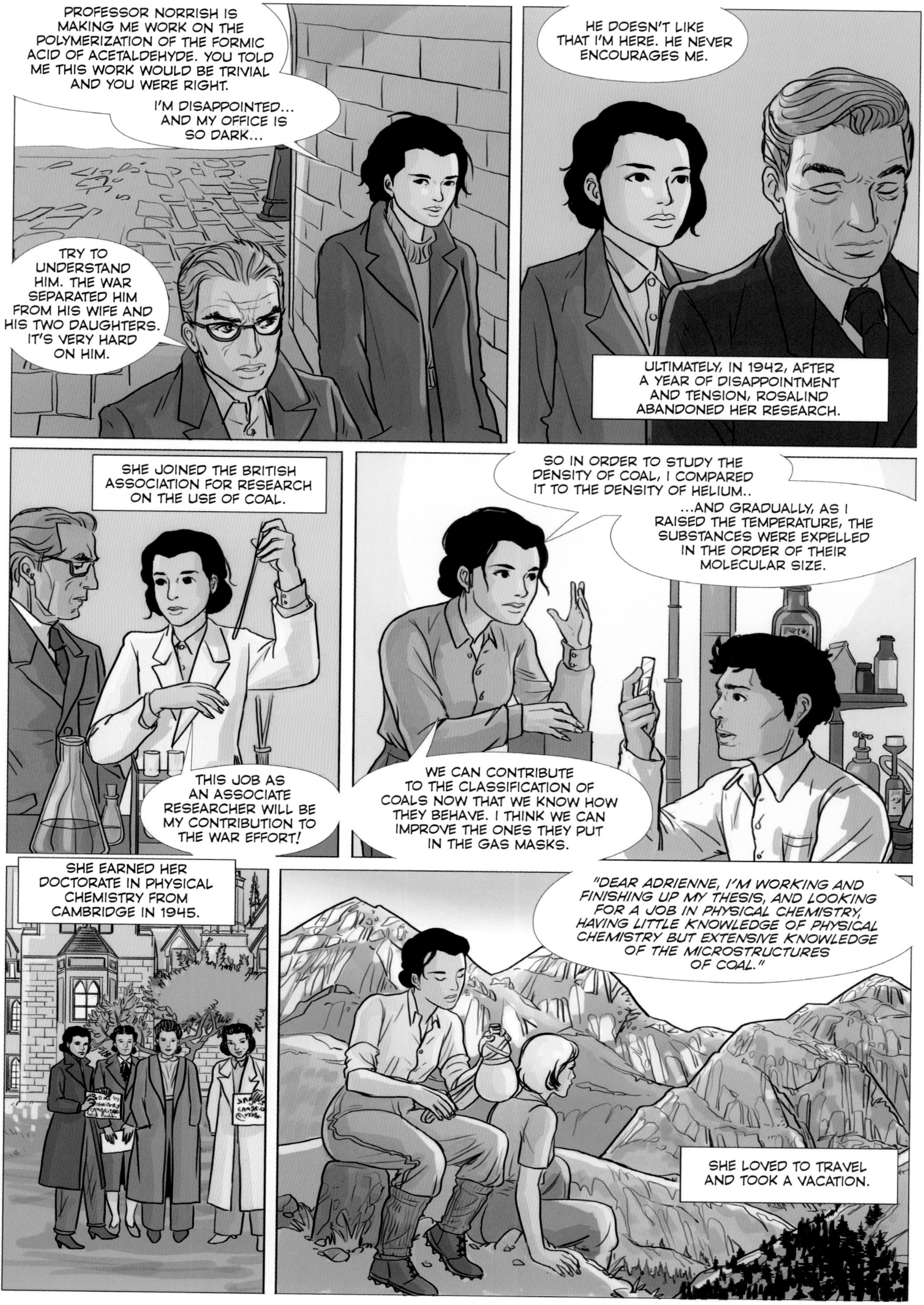
PROFESSOR NORRISH IS MAKING ME WORK ON THE POLYMERIZATION OF THE FORMIC ACID OF ACETALDEHYDE. YOU TOLD ME THIS WORK WOULD BE TRIVIAL AND YOU WERE RIGHT.
I'M DISAPPOINTED… AND MY OFFICE IS SO DARK…
TRY TO UNDERSTAND HIM. THE WAR SEPARATED HIM FROM HIS WIFE AND HIS TWO DAUGHTERS. IT'S VERY HARD ON HIM.
HE DOESN'T LIKE THAT I'M HERE. HE NEVER ENCOURAGES ME.
ULTIMATELY, IN 1942, AFTER A YEAR OF DISAPPOINTMENT AND TENSION, ROSALIND ABANDONED HER RESEARCH.
SHE JOINED THE BRITISH ASSOCIATION FOR RESEARCH ON THE USE OF COAL.
THIS JOB AS AN ASSOCIATE RESEARCHER WILL BE MY CONTRIBUTION TO THE WAR EFFORT!
SO IN ORDER TO STUDY THE DENSITY OF COAL, I COMPARED IT TO THE DENSITY OF HELIUM..
…AND GRADUALLY, AS I RAISED THE TEMPERATURE, THE SUBSTANCES WERE EXPELLED IN THE ORDER OF THEIR MOLECULAR SIZE.
WE CAN CONTRIBUTE TO THE CLASSIFICATION OF COALS NOW THAT WE KNOW HOW THEY BEHAVE. I THINK WE CAN IMPROVE THE ONES THEY PUT IN THE GAS MASKS.
SHE EARNED HER DOCTORATE IN PHYSICAL CHEMISTRY FROM CAMBRIDGE IN 1945.
"DEAR ADRIENNE, I'M WORKING AND FINISHING UP MY THESIS, AND LOOKING FOR A JOB IN PHYSICAL CHEMISTRY, HAVING LITTLE KNOWLEDGE OF PHYSICAL CHEMISTRY BUT EXTENSIVE KNOWLEDGE OF THE MICROSTRUCTURES OF COAL."
SHE LOVED TO TRAVEL AND TOOK A VACATION.

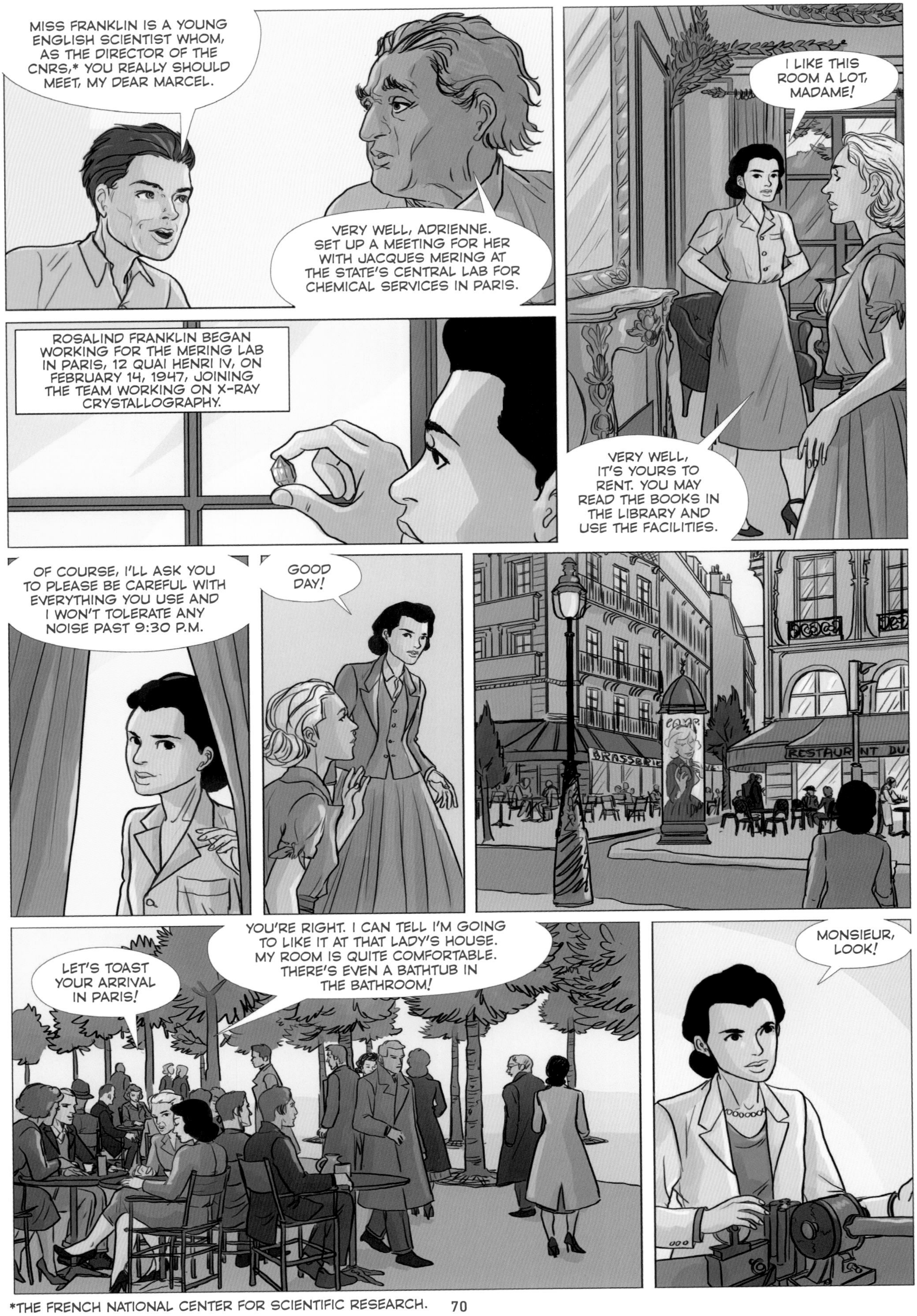
MISS FRANKLIN IS A YOUNG ENGLISH SCIENTIST WHOM, AS THE DIRECTOR OF THE CNRS,* YOU REALLY SHOULD MEET, MY DEAR MARCEL.
VERY WELL, ADRIENNE. SET UP A MEETING FOR HER WITH JACQUES MERING AT THE STATE'S CENTRAL LAB FOR CHEMICAL SERVICES IN PARIS.
I LIKE THIS ROOM A LOT, MADAME!
ROSALIND FRANKLIN BEGAN WORKING FOR THE MERING LAB IN PARIS, 12 QUAI HENRI IV, ON FEBRUARY 14, 1947, JOINING THE TEAM WORKING ON X-RAY CRYSTALLOGRAPHY.
VERY WELL, IT'S YOURS TO RENT. YOU MAY READ THE BOOKS IN THE LIBRARY AND USE THE FACILITIES.
OF COURSE, I'LL ASK YOU TO PLEASE BE CAREFUL WITH EVERYTHING YOU USE AND I WON'T TOLERATE ANY NOISE PAST 9:30 P.M.
GOOD DAY!
BRASSERIE
RESTAURANT DU
LET'S TOAST YOUR ARRIVAL IN PARIS!
YOU'RE RIGHT. I CAN TELL I'M GOING TO LIKE IT AT THAT LADY'S HOUSE. MY ROOM IS QUITE COMFORTABLE. THERE'S EVEN A BATHTUB IN THE BATHROOM!
MONSIEUR, LOOK!

*THE FRENCH NATIONAL CENTER FOR SCIENTIFIC RESEARCH.

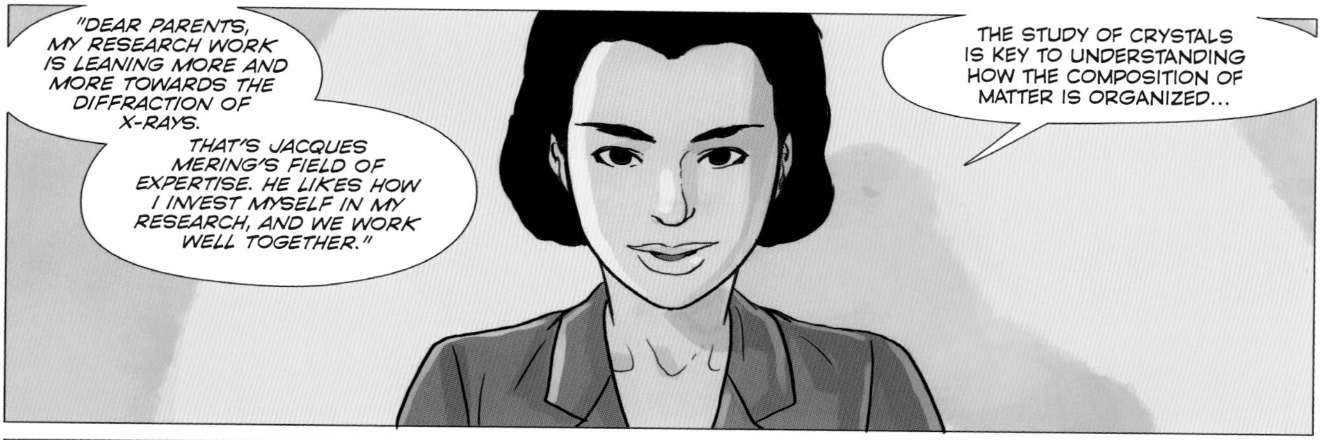
"DEAR PARENTS, MY RESEARCH WORK IS LEANING MORE AND MORE TOWARDS THE DIFFRACTION OF X-RAYS.
THAT'S JACQUES MERING'S FIELD OF EXPERTISE. HE LIKES HOW I INVEST MYSELF IN MY RESEARCH, AND WE WORK WELL TOGETHER."
THE STUDY OF CRYSTALS IS KEY TO UNDERSTANDING HOW THE COMPOSITION OF MATTER IS ORGANIZED...

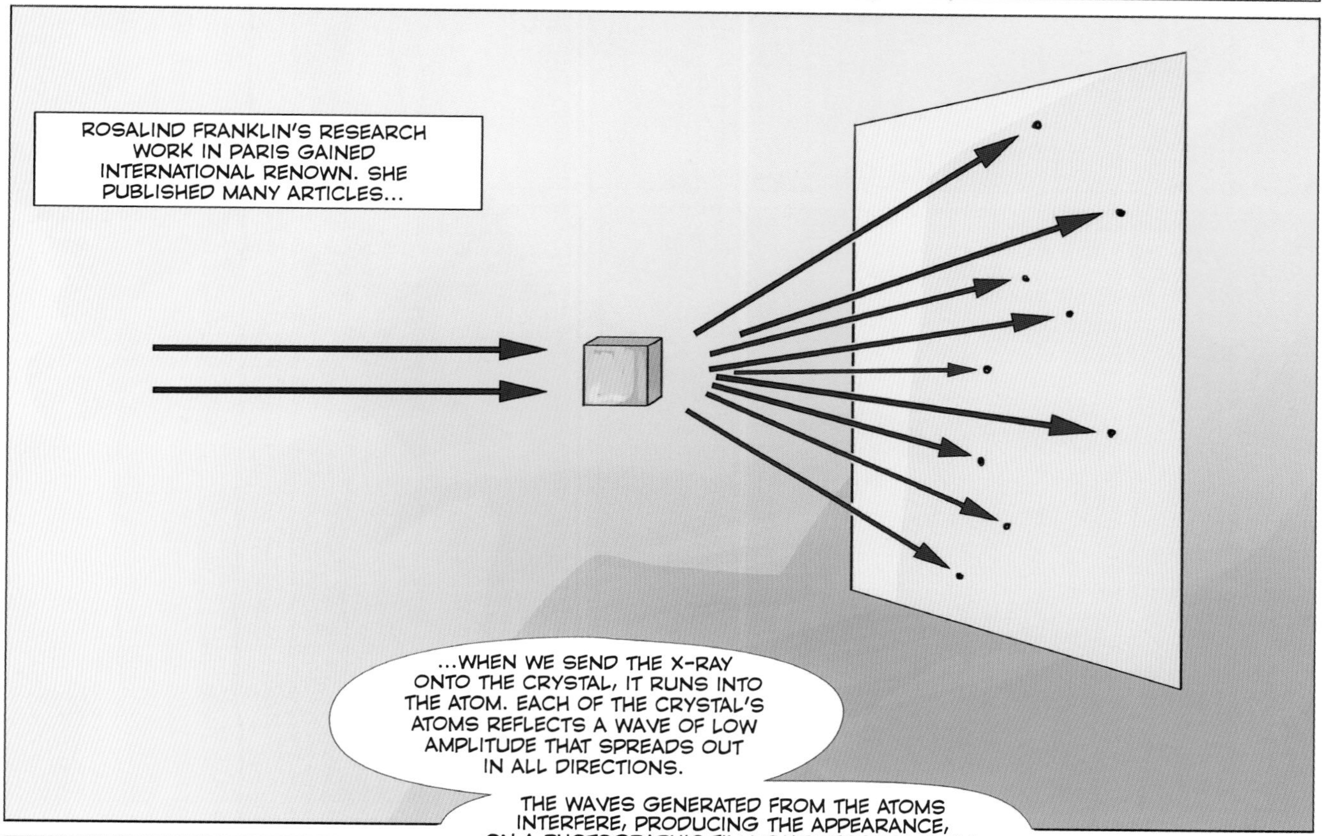
ROSALIND FRANKLIN'S RESEARCH WORK IN PARIS GAINED INTERNATIONAL RENOWN. SHE PUBLISHED MANY ARTICLES...
...WHEN WE SEND THE X-RAY ONTO THE CRYSTAL, IT RUNS INTO THE ATOM. EACH OF THE CRYSTAL'S ATOMS REFLECTS A WAVE OF LOW AMPLITUDE THAT SPREADS OUT IN ALL DIRECTIONS.
THE WAVES GENERATED FROM THE ATOMS INTERFERE, PRODUCING THE APPEARANCE, ON A PHOTOGRAPHIC FILM THAT RECEIVES THEM, OF SPOTS THAT CORRESPOND TO THE MAXIMUM OF THE WAVES IN PHASE.

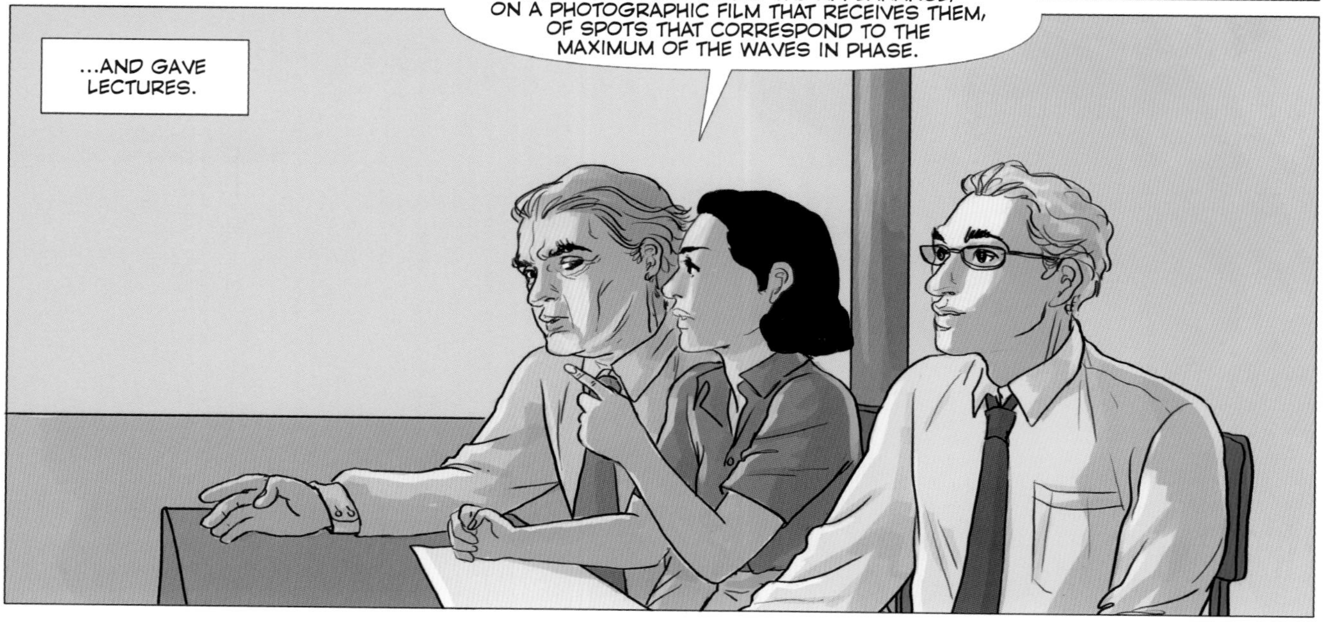
...AND GAVE LECTURES.

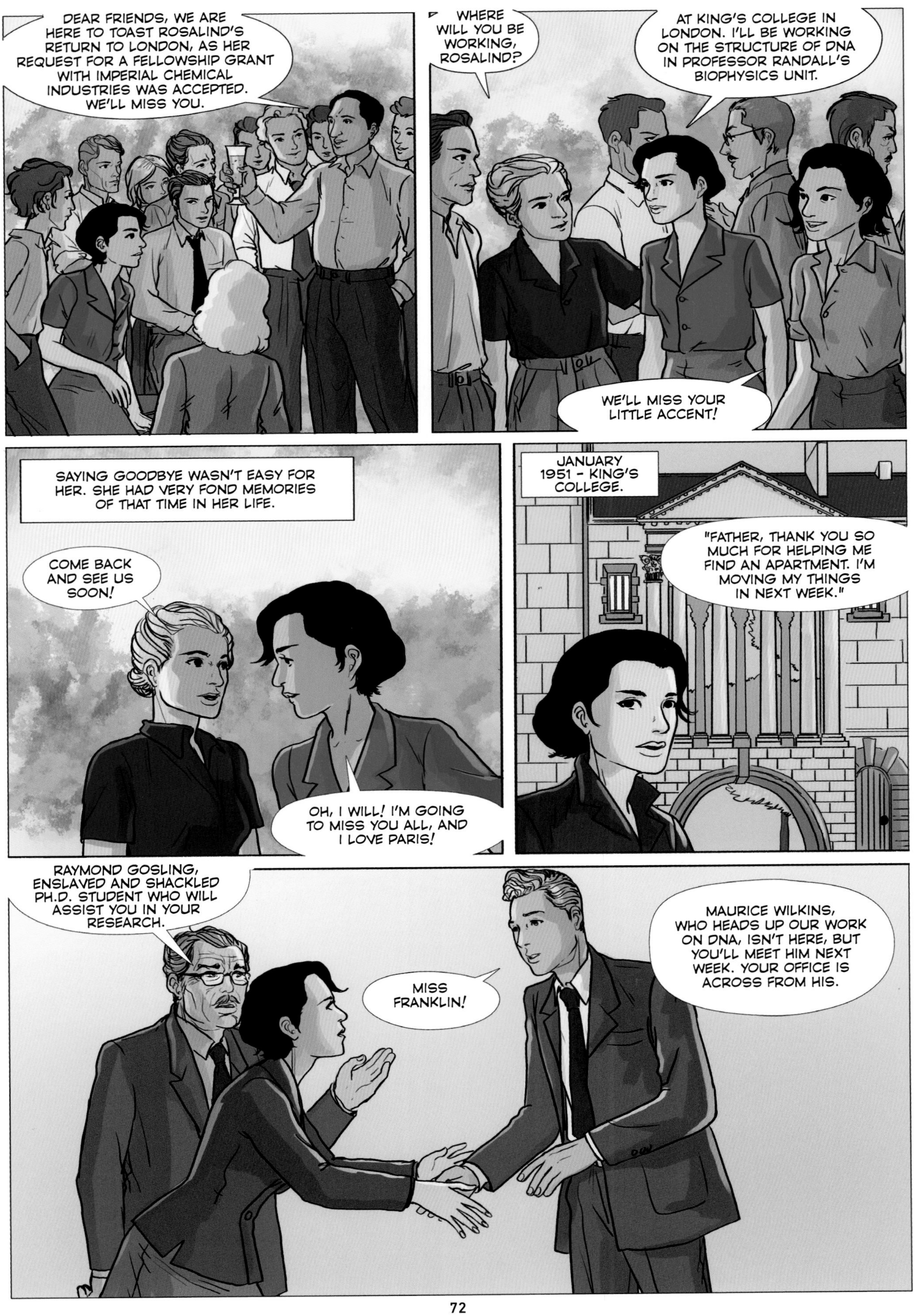
DEAR FRIENDS, WE ARE HERE TO TOAST ROSALIND'S RETURN TO LONDON, AS HER REQUEST FOR A FELLOWSHIP GRANT WITH IMPERIAL CHEMICAL INDUSTRIES WAS ACCEPTED. WE'LL MISS YOU.
WHERE WILL YOU BE WORKING, ROSALIND?
AT KING'S COLLEGE IN LONDON. I'LL BE WORKING ON THE STRUCTURE OF DNA IN PROFESSOR RANDALL'S BIOPHYSICS UNIT.
WE'LL MISS YOUR LITTLE ACCENT!
SAYING GOODBYE WASN'T EASY FOR HER. SHE HAD VERY FOND MEMORIES OF THAT TIME IN HER LIFE.
COME BACK AND SEE US SOON!
OH, I WILL! I'M GOING TO MISS YOU ALL, AND I LOVE PARIS!
JANUARY 1951 - KING'S COLLEGE.
"FATHER, THANK YOU SO MUCH FOR HELPING ME FIND AN APARTMENT. I'M MOVING MY THINGS IN NEXT WEEK."
RAYMOND GOSLING, ENSLAVED AND SHACKLED PH.D. STUDENT WHO WILL ASSIST YOU IN YOUR RESEARCH.
MISS FRANKLIN!
MAURICE WILKINS, WHO HEADS UP OUR WORK ON DNA, ISN'T HERE, BUT YOU'LL MEET HIM NEXT WEEK. YOUR OFFICE IS ACROSS FROM HIS.

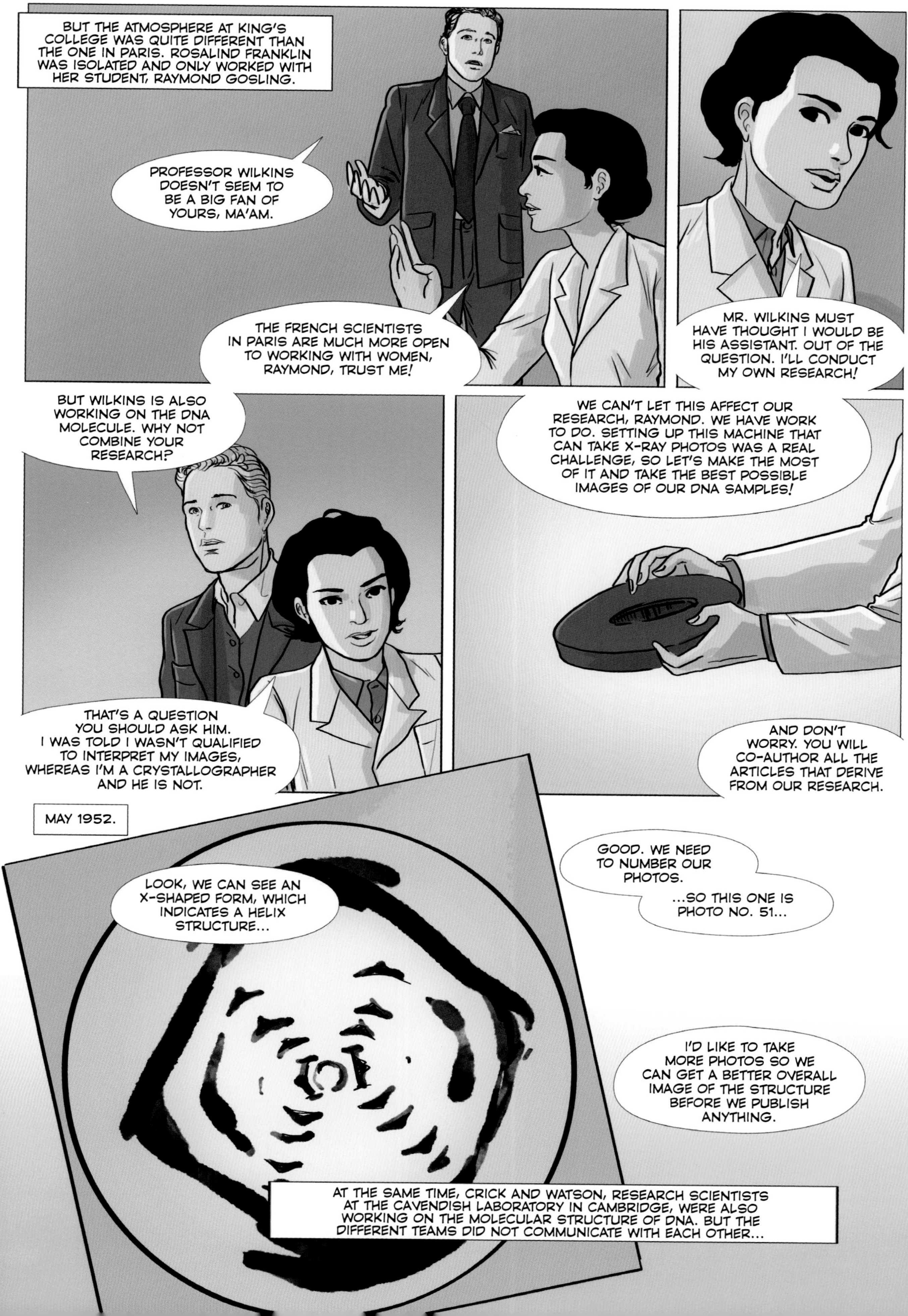
BUT THE ATMOSPHERE AT KING'S COLLEGE WAS QUITE DIFFERENT THAN THE ONE IN PARIS. ROSALIND FRANKLIN WAS ISOLATED AND ONLY WORKED WITH HER STUDENT, RAYMOND GOSLING.
PROFESSOR WILKINS DOESN'T SEEM TO BE A BIG FAN OF YOURS, MA'AM.
THE FRENCH SCIENTISTS IN PARIS ARE MUCH MORE OPEN TO WORKING WITH WOMEN, RAYMOND, TRUST ME!
MR. WILKINS MUST HAVE THOUGHT I WOULD BE HIS ASSISTANT. OUT OF THE QUESTION. I'LL CONDUCT MY OWN RESEARCH!
BUT WILKINS IS ALSO WORKING ON THE DNA MOLECULE. WHY NOT COMBINE YOUR RESEARCH?
THAT'S A QUESTION YOU SHOULD ASK HIM. I WAS TOLD I WASN'T QUALIFIED TO INTERPRET MY IMAGES, WHEREAS I'M A CRYSTALLOGRAPHER AND HE IS NOT.
WE CAN'T LET THIS AFFECT OUR RESEARCH, RAYMOND. WE HAVE WORK TO DO. SETTING UP THIS MACHINE THAT CAN TAKE X-RAY PHOTOS WAS A REAL CHALLENGE, SO LET'S MAKE THE MOST OF IT AND TAKE THE BEST POSSIBLE IMAGES OF OUR DNA SAMPLES!
AND DON'T WORRY. YOU WILL CO-AUTHOR ALL THE ARTICLES THAT DERIVE FROM OUR RESEARCH.
MAY 1952.
LOOK, WE CAN SEE AN X-SHAPED FORM, WHICH INDICATES A HELIX STRUCTURE...
GOOD. WE NEED TO NUMBER OUR PHOTOS.
...SO THIS ONE IS PHOTO NO. 51...
I'D LIKE TO TAKE MORE PHOTOS SO WE CAN GET A BETTER OVERALL IMAGE OF THE STRUCTURE BEFORE WE PUBLISH ANYTHING.
AT THE SAME TIME, CRICK AND WATSON, RESEARCH SCIENTISTS AT THE CAVENDISH LABORATORY IN CAMBRIDGE, WERE ALSO WORKING ON THE MOLECULAR STRUCTURE OF DNA. BUT THE DIFFERENT TEAMS DID NOT COMMUNICATE WITH EACH OTHER...

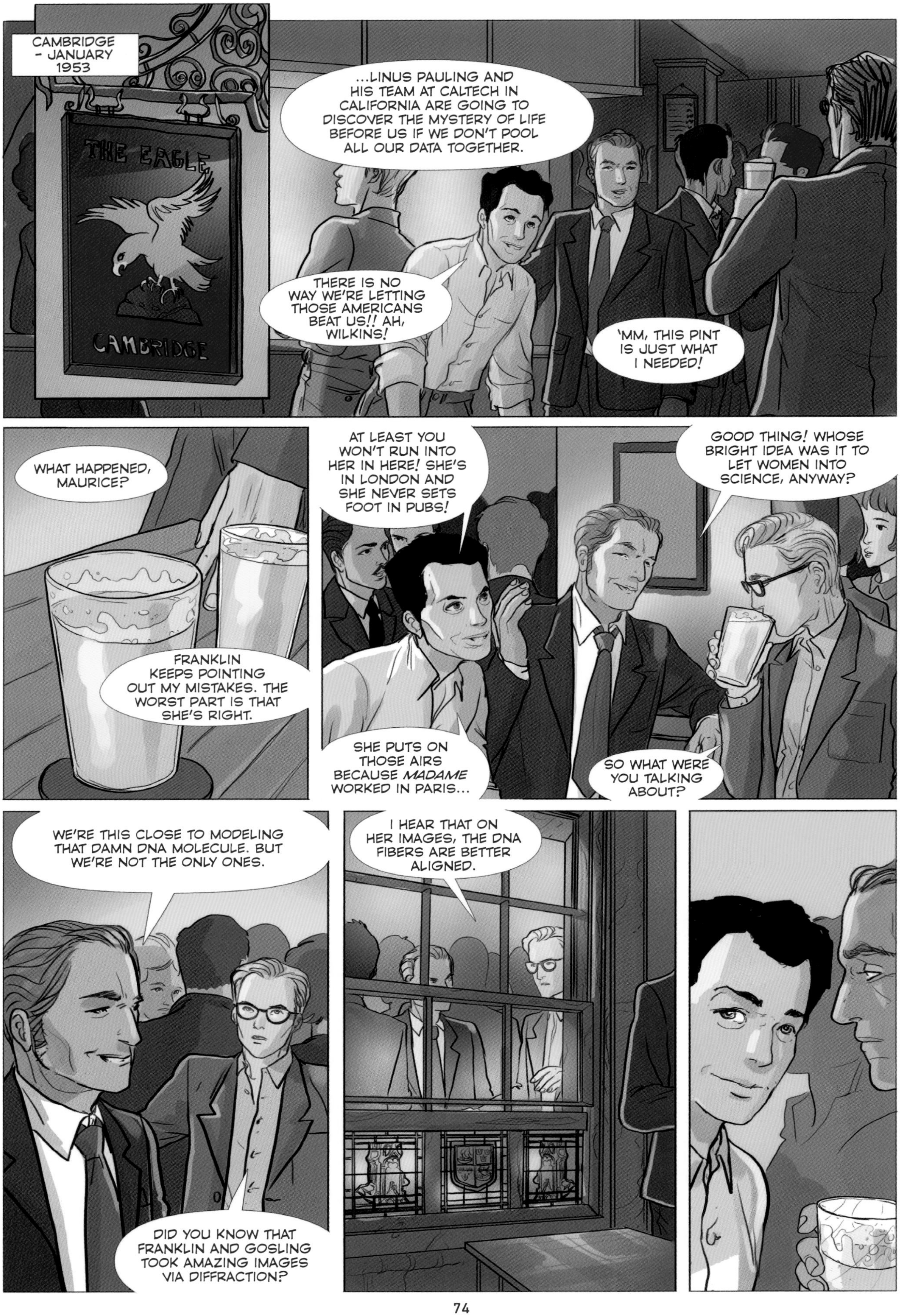
CAMBRIDGE - JANUARY 1953
THE EAGLE
CAMBRIDGE
...LINUS PAULING AND HIS TEAM AT CALTECH IN CALIFORNIA ARE GOING TO DISCOVER THE MYSTERY OF LIFE BEFORE US IF WE DON'T POOL ALL OUR DATA TOGETHER.
THERE IS NO WAY WE'RE LETTING THOSE AMERICANS BEAT US!! AH, WILKINS!
'MM, THIS PINT IS JUST WHAT I NEEDED!
WHAT HAPPENED, MAURICE?
FRANKLIN KEEPS POINTING OUT MY MISTAKES. THE WORST PART IS THAT SHE'S RIGHT.
AT LEAST YOU WON'T RUN INTO HER IN HERE! SHE'S IN LONDON AND SHE NEVER SETS FOOT IN PUBS!
GOOD THING! WHOSE BRIGHT IDEA WAS IT TO LET WOMEN INTO SCIENCE, ANYWAY?
SHE PUTS ON THOSE AIRS BECAUSE *MADAME* WORKED IN PARIS...
SO WHAT WERE YOU TALKING ABOUT?
WE'RE THIS CLOSE TO MODELING THAT DAMN DNA MOLECULE. BUT WE'RE NOT THE ONLY ONES.
DID YOU KNOW THAT FRANKLIN AND GOSLING TOOK AMAZING IMAGES VIA DIFFRACTION?
I HEAR THAT ON HER IMAGES, THE DNA FIBERS ARE BETTER ALIGNED.

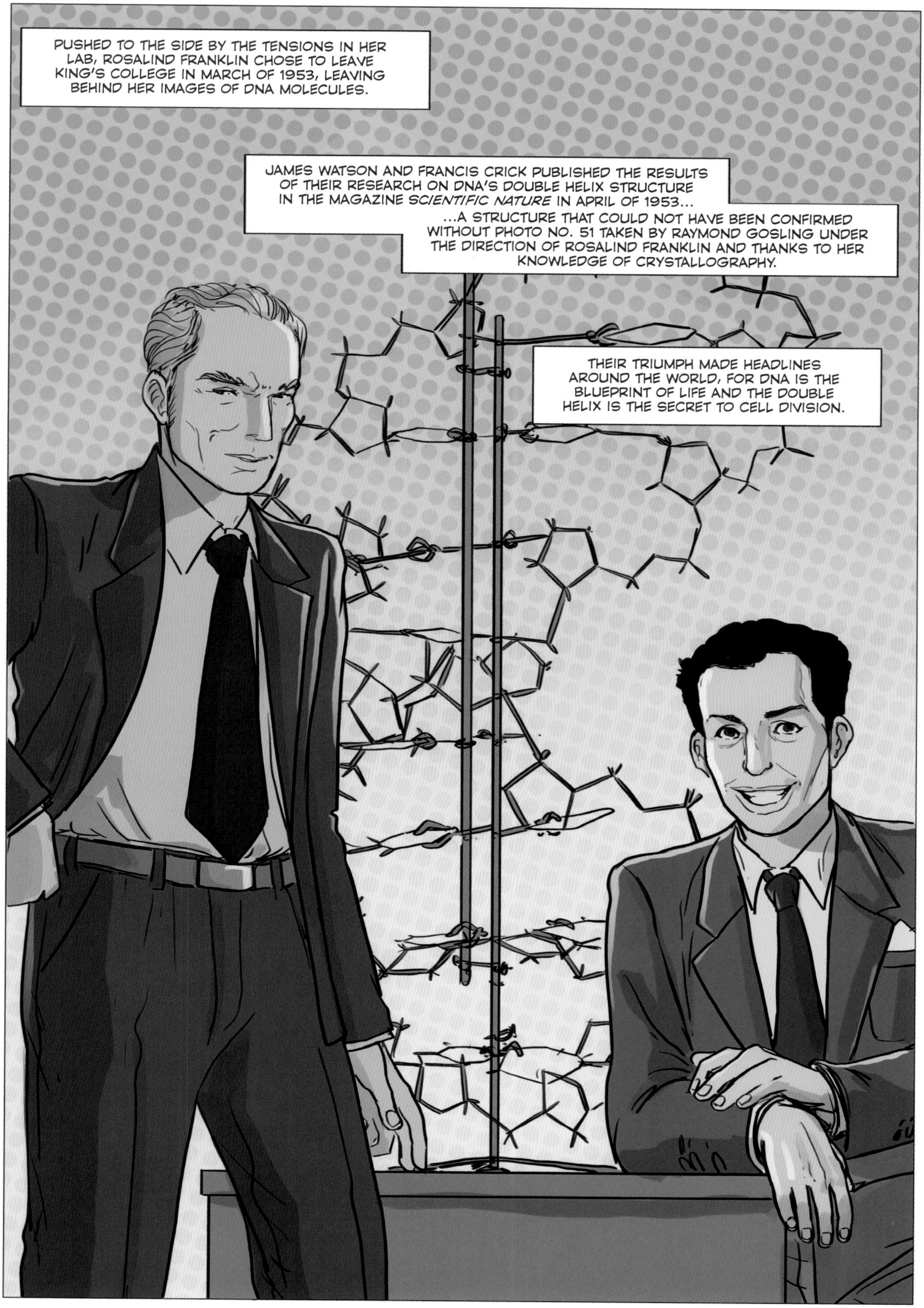
PUSHED TO THE SIDE BY THE TENSIONS IN HER LAB, ROSALIND FRANKLIN CHOSE TO LEAVE KING'S COLLEGE IN MARCH OF 1953, LEAVING BEHIND HER IMAGES OF DNA MOLECULES.
JAMES WATSON AND FRANCIS CRICK PUBLISHED THE RESULTS OF THEIR RESEARCH ON DNA'S DOUBLE HELIX STRUCTURE IN THE MAGAZINE *SCIENTIFIC NATURE* IN APRIL OF 1953...
...A STRUCTURE THAT COULD NOT HAVE BEEN CONFIRMED WITHOUT PHOTO NO. 51 TAKEN BY RAYMOND GOSLING UNDER THE DIRECTION OF ROSALIND FRANKLIN AND THANKS TO HER KNOWLEDGE OF CRYSTALLOGRAPHY.
THEIR TRIUMPH MADE HEADLINES AROUND THE WORLD, FOR DNA IS THE BLUEPRINT OF LIFE AND THE DOUBLE HELIX IS THE SECRET TO CELL DIVISION.

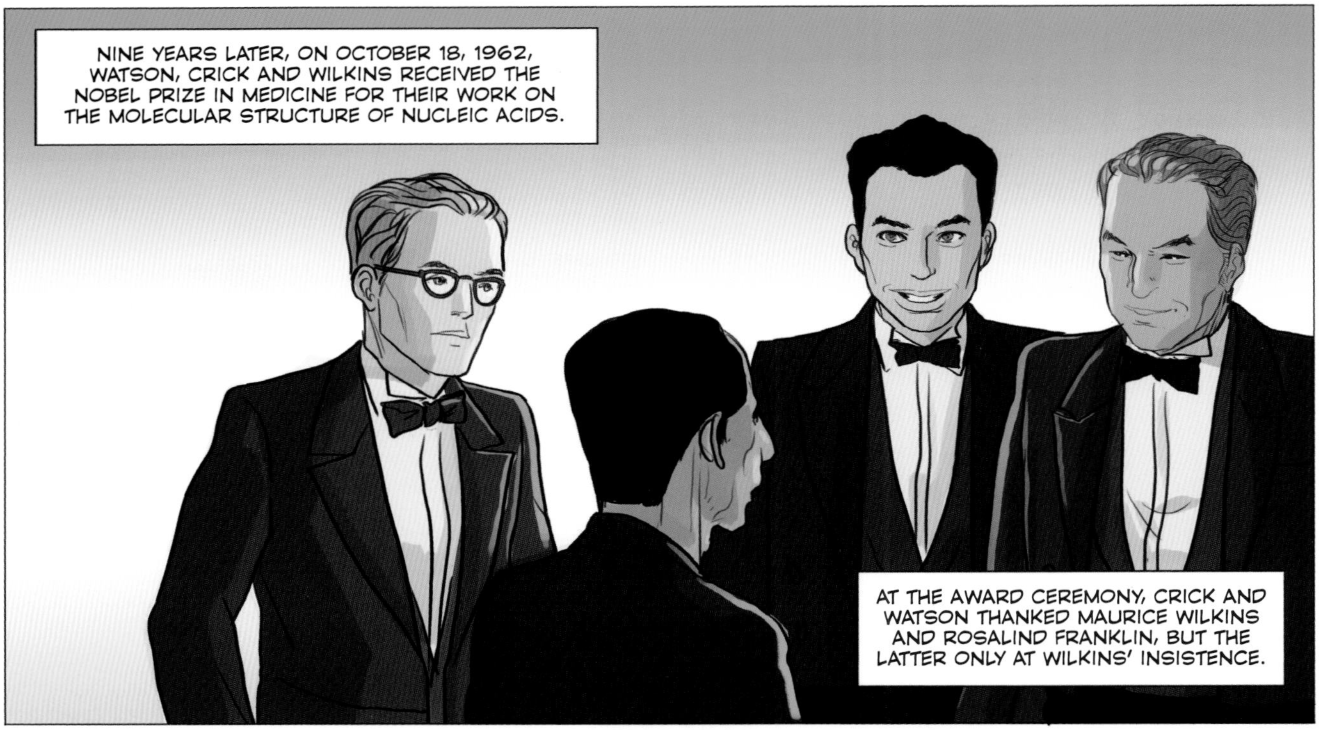
NINE YEARS LATER, ON OCTOBER 18, 1962, WATSON, CRICK AND WILKINS RECEIVED THE NOBEL PRIZE IN MEDICINE FOR THEIR WORK ON THE MOLECULAR STRUCTURE OF NUCLEIC ACIDS.
AT THE AWARD CEREMONY, CRICK AND WATSON THANKED MAURICE WILKINS AND ROSALIND FRANKLIN, BUT THE LATTER ONLY AT WILKINS' INSISTENCE.

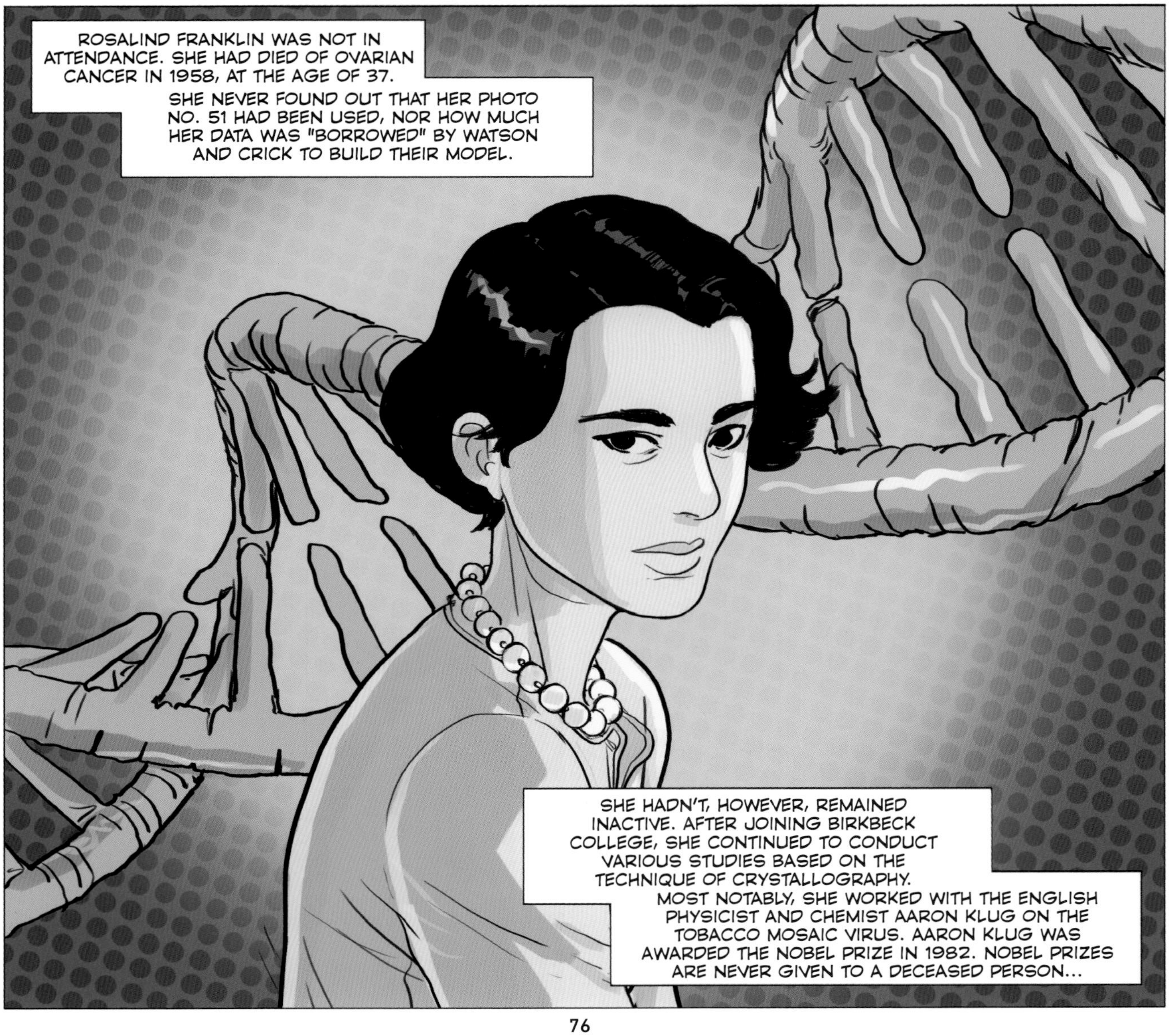
ROSALIND FRANKLIN WAS NOT IN ATTENDANCE. SHE HAD DIED OF OVARIAN CANCER IN 1958, AT THE AGE OF 37.
SHE NEVER FOUND OUT THAT HER PHOTO NO. 51 HAD BEEN USED, NOR HOW MUCH HER DATA WAS "BORROWED" BY WATSON AND CRICK TO BUILD THEIR MODEL.
SHE HADN'T, HOWEVER, REMAINED INACTIVE. AFTER JOINING BIRKBECK COLLEGE, SHE CONTINUED TO CONDUCT VARIOUS STUDIES BASED ON THE TECHNIQUE OF CRYSTALLOGRAPHY.
MOST NOTABLY, SHE WORKED WITH THE ENGLISH PHYSICIST AND CHEMIST AARON KLUG ON THE TOBACCO MOSAIC VIRUS. AARON KLUG WAS AWARDED THE NOBEL PRIZE IN 1982. NOBEL PRIZES ARE NEVER GIVEN TO A DECEASED PERSON...

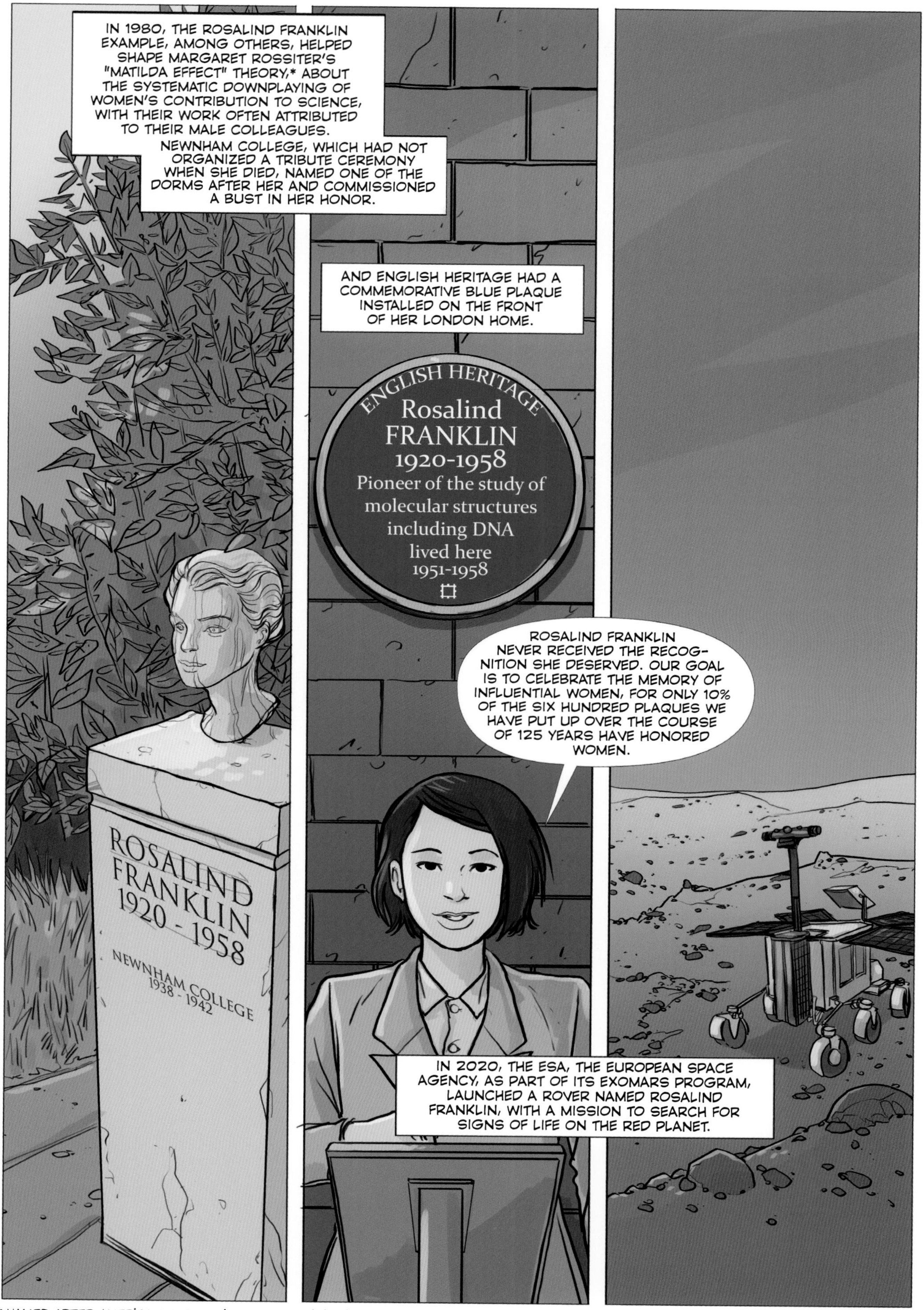

IN 1980, THE ROSALIND FRANKLIN EXAMPLE, AMONG OTHERS, HELPED SHAPE MARGARET ROSSITER'S "MATILDA EFFECT" THEORY,* ABOUT THE SYSTEMATIC DOWNPLAYING OF WOMEN'S CONTRIBUTION TO SCIENCE, WITH THEIR WORK OFTEN ATTRIBUTED TO THEIR MALE COLLEAGUES.
NEWNHAM COLLEGE, WHICH HAD NOT ORGANIZED A TRIBUTE CEREMONY WHEN SHE DIED, NAMED ONE OF THE DORMS AFTER HER AND COMMISSIONED A BUST IN HER HONOR.
ROSALIND FRANKLIN 1920 - 1958
NEWNHAM COLLEGE 1938 - 1942
AND ENGLISH HERITAGE HAD A COMMEMORATIVE BLUE PLAQUE INSTALLED ON THE FRONT OF HER LONDON HOME.
ENGLISH HERITAGE Rosalind FRANKLIN 1920-1958 Pioneer of the study of molecular structures including DNA lived here 1951-1958
ROSALIND FRANKLIN NEVER RECEIVED THE RECOGNITION SHE DESERVED. OUR GOAL IS TO CELEBRATE THE MEMORY OF INFLUENTIAL WOMEN, FOR ONLY 10% OF THE SIX HUNDRED PLAQUES WE HAVE PUT UP OVER THE COURSE OF 125 YEARS HAVE HONORED WOMEN.
IN 2020, THE ESA, THE EUROPEAN SPACE AGENCY, AS PART OF ITS EXOMARS PROGRAM, LAUNCHED A ROVER NAMED ROSALIND FRANKLIN, WITH A MISSION TO SEARCH FOR SIGNS OF LIFE ON THE RED PLANET.

* NAMED AFTER AMERICAN SUFFRAGIST AND ABOLITIONIST MATILDA JOSLYN GAGE.

THE END

Sophie Germain

1776 - 1831

French mathematician, philosopher and physicist.

Sophie Germain was born in Paris in 1776 and learned mathematics on her own by reading the many books in the family library. Despite initial parental disapproval, she refused to get discouraged, and her obstinacy eventually won over her parents, who supported her when she was accepted at the recently founded Polytechnic School by pretending to be a boy named Antoine-Auguste Le Blanc.

It was under that identity that she corresponded with Joseph-Louis Lagrange (a mathematician, mechanic and astronomer), who eventually discovered the subterfuge. Impressed, he too decided to offer her his support.

She began working on Fermat's theorem, performed a demonstration on prime numbers and a series of experiments on the vibrations of elastic surfaces. She wrote three papers comparing the experiment and the mathematical theory of elastic surfaces. She became the first woman to win the Grand Prize for Mathematical Sciences from the Academy of Sciences.

Despite all these accomplishments, she was never accepted as an equal of men and she felt alone and foreign in the scientific community. She died at the age of 55 before receiving an honorary doctorate from by the University of Gottingen.

Irène Joliot-Curie

1897 – 1956

French chemist and physicist
Nobel Prize in Chemistry 1935
(shared with Frédéric Joliot)

The daughter of illustrious scientists Pierre and Marie Curie, **Irène Joliot-Curie** was born in Paris in 1897. Immersed early on in a family atmosphere steeped in science and research, Irene quite naturally followed in her parents' footsteps and, after obtaining her baccalaureate, enlisted in the war as a radiologist nurse and surgical assistant.

After World War I ended, she became her mother's assistant at the Radium Institute in Paris.

There, she met another research scientist, Frederic Joliot, whom she fell in love with and married. They had two children together and also shared the 1935 Nobel Prize in chemistry for their discovery of artificial radioactivity.

Irène Joliot-Curie was also a dedicated advocate for women's rights and was appointed to the post of secretary of state for scientific research under the left-wing Front Populaire government in June 1936, even though women didn't yet have the right to vote.

A professor and the director of the Radium Institute, she died at the age of 58, likely as a result of the significant amount of radiation to which she was exposed.

Jocelyn Bell Burnell

1943 –

Irish astrophysicist
Breakthrough Prize 2018

Jocelyn Bell Burnell was born in 1943 in Belfast, Northern Ireland. She had a strong interest in astronomy as a child, which was only reinforced when she accompanied her father to the Armagh Planetarium, where he worked.

While a student at Cambridge University, she helped build a radio telescope and used it every day until one bright spot intrigued her enough to write down data and analyze the findings, which, in 1967, led to her discovery of pulsars, stellar objects that produce a periodic signal.

This was such a monumental discovery in the field of astronomy that Anthony Hewish, her research supervisor as well as her collaborator, was awarded the Nobel Prize in Physics in 1974.

Jocelyn, who had continued her research despite Hewish's doubts, was not nominated.

But 51 years after her discovery, in 2018, she finally received the recognition she deserved. She was awarded the Breakthrough Prize in Fundamental Physics with an endowment of three million dollars in recognition of her work and her influence on the scientific community.

She used the money for a scholarship fund for minority students to study advanced physics at Oxford.

MAE JEMISON

1956 -
American Astronaut
Buzz Aldrin's Space Pioneers Award - 2017

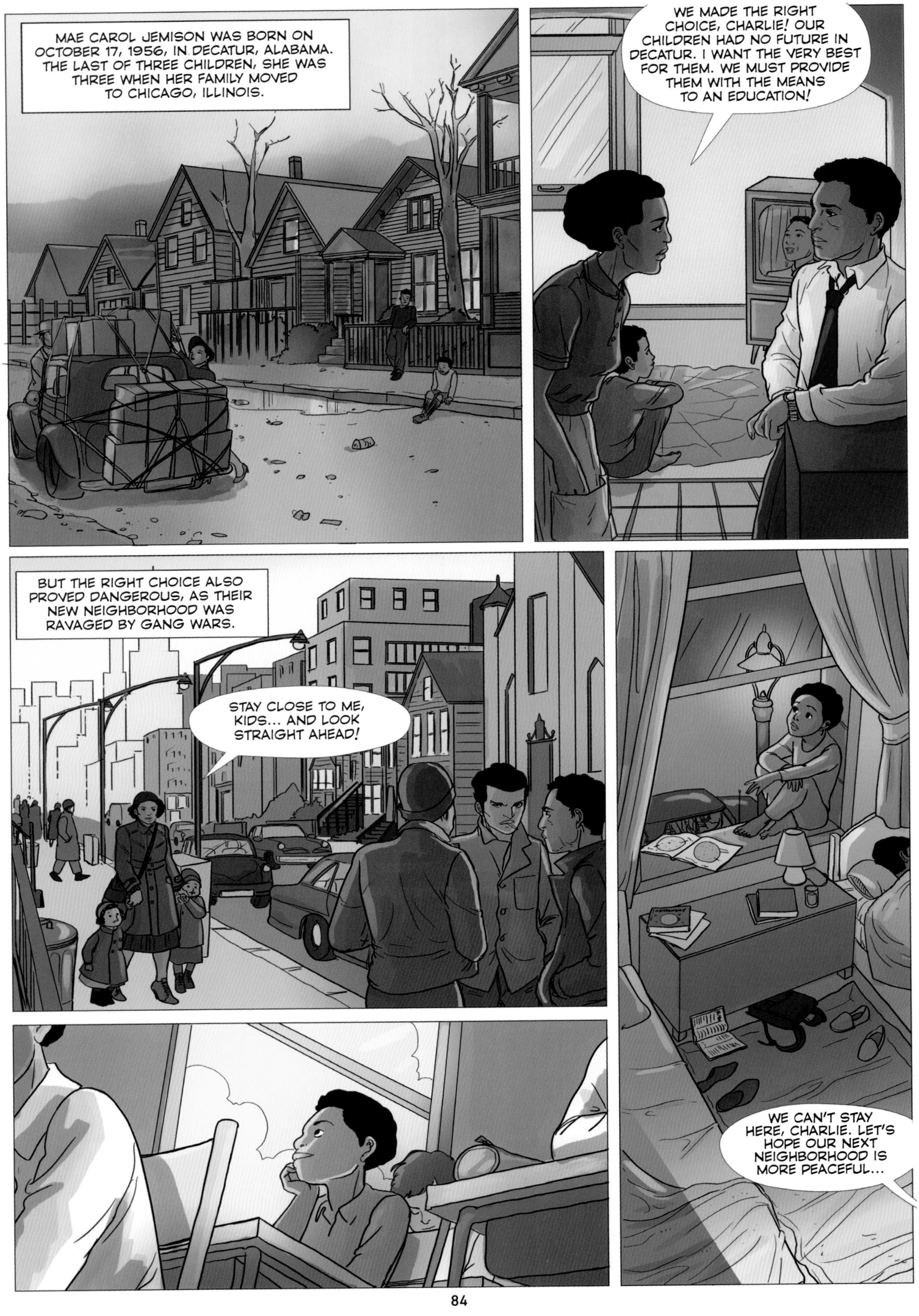
MAE CAROL JEMISON WAS BORN ON OCTOBER 17, 1956, IN DECATUR, ALABAMA. THE LAST OF THREE CHILDREN, SHE WAS THREE WHEN HER FAMILY MOVED TO CHICAGO, ILLINOIS.
WE MADE THE RIGHT CHOICE, CHARLIE! OUR CHILDREN HAD NO FUTURE IN DECATUR. I WANT THE VERY BEST FOR THEM. WE MUST PROVIDE THEM WITH THE MEANS TO AN EDUCATION!
BUT THE RIGHT CHOICE ALSO PROVED DANGEROUS, AS THEIR NEW NEIGHBORHOOD WAS RAVAGED BY GANG WARS.
STAY CLOSE TO ME, KIDS... AND LOOK STRAIGHT AHEAD!
WE CAN'T STAY HERE, CHARLIE. LET'S HOPE OUR NEXT NEIGHBORHOOD IS MORE PEACEFUL...

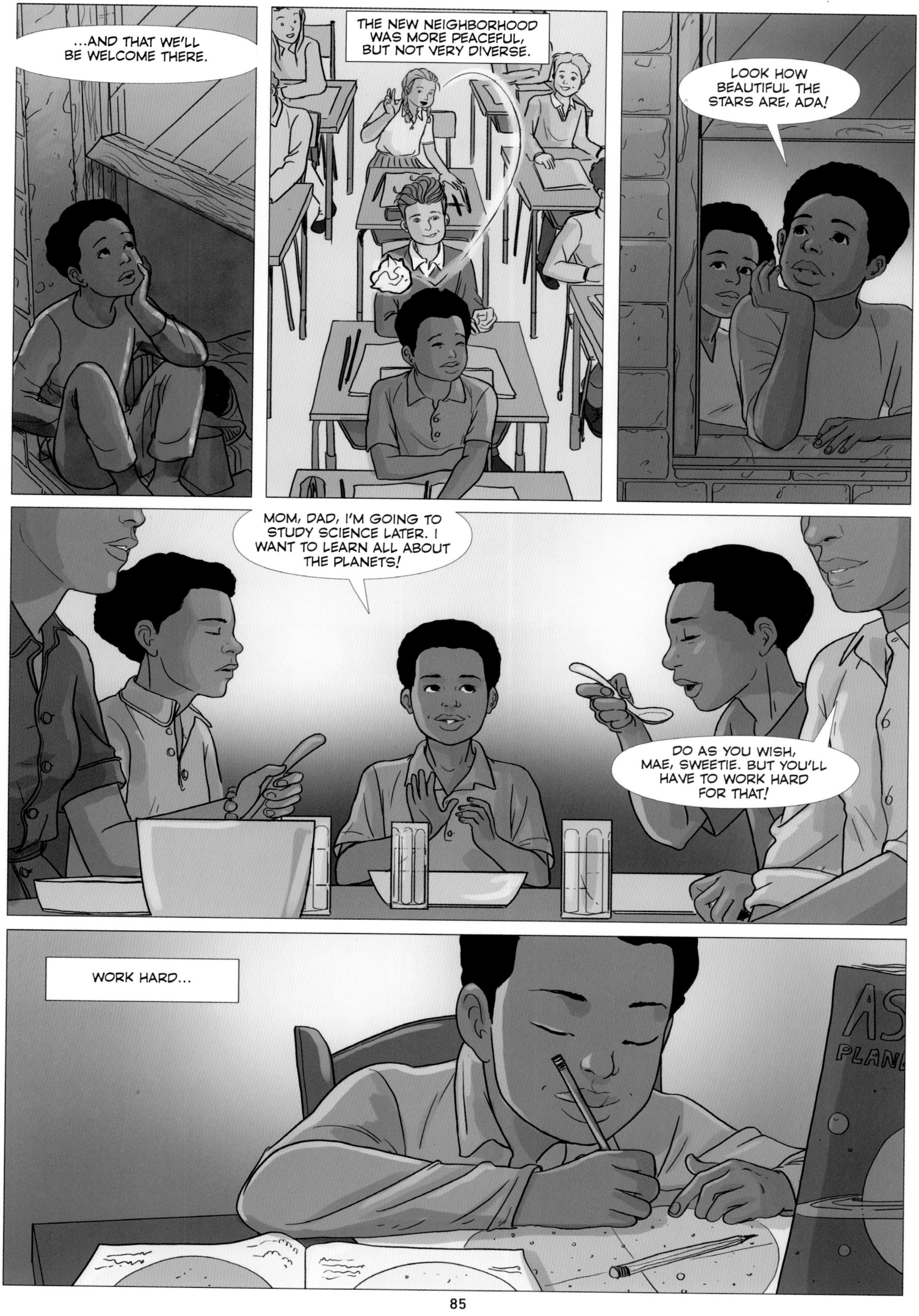
...AND THAT WE'LL BE WELCOME THERE.
THE NEW NEIGHBORHOOD WAS MORE PEACEFUL, BUT NOT VERY DIVERSE.
LOOK HOW BEAUTIFUL THE STARS ARE, ADA!
MOM, DAD, I'M GOING TO STUDY SCIENCE LATER. I WANT TO LEARN ALL ABOUT THE PLANETS!
DO AS YOU WISH, MAE, SWEETIE. BUT YOU'LL HAVE TO WORK HARD FOR THAT!
WORK HARD...

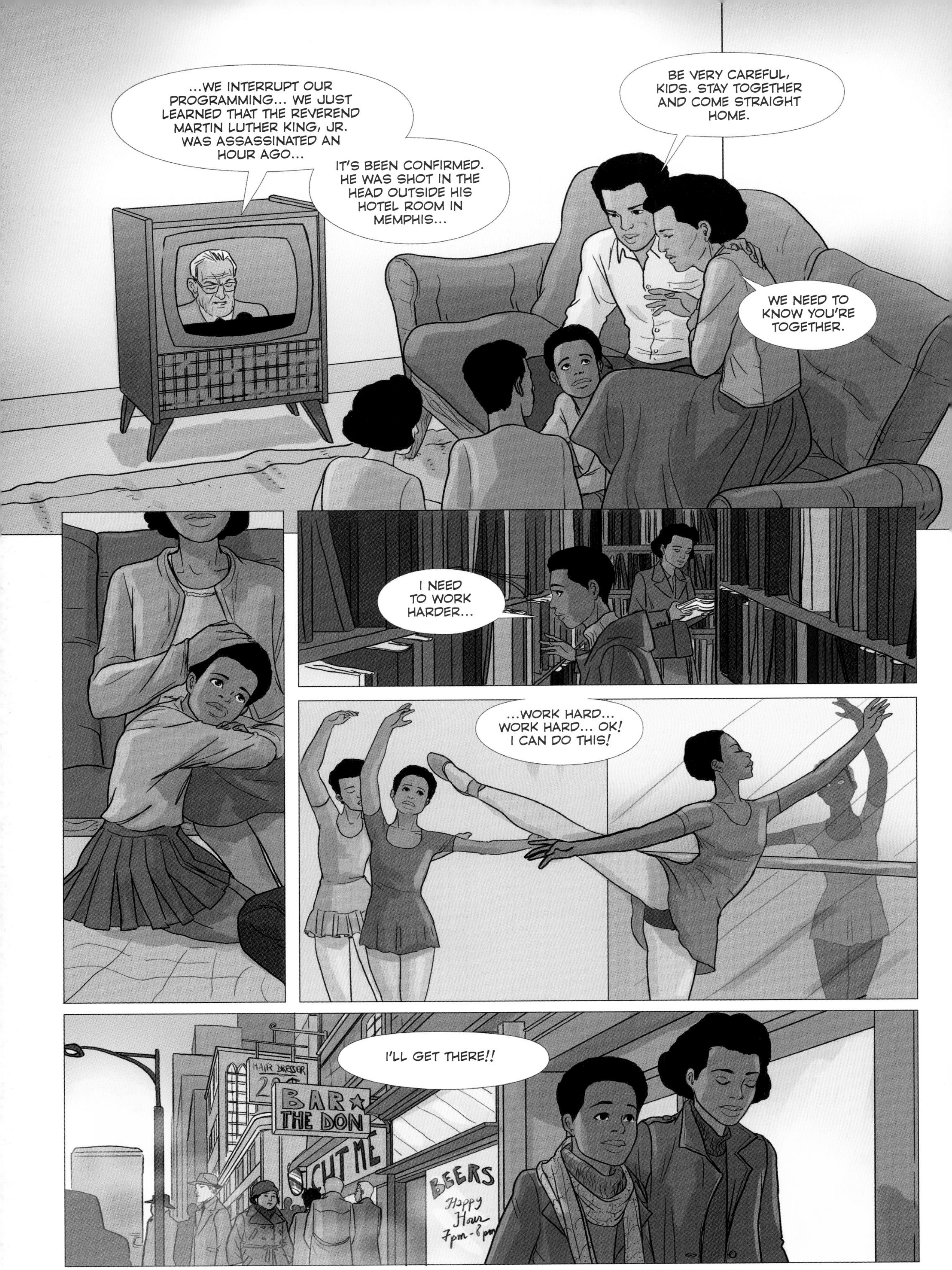
...WE INTERRUPT OUR PROGRAMMING... WE JUST LEARNED THAT THE REVEREND MARTIN LUTHER KING, JR. WAS ASSASSINATED AN HOUR AGO...
IT'S BEEN CONFIRMED. HE WAS SHOT IN THE HEAD OUTSIDE HIS HOTEL ROOM IN MEMPHIS...
BE VERY CAREFUL, KIDS. STAY TOGETHER AND COME STRAIGHT HOME.
WE NEED TO KNOW YOU'RE TOGETHER.
I NEED TO WORK HARDER...
...WORK HARD... WORK HARD... OK! I CAN DO THIS!
I'LL GET THERE!!
HAIR DRESSER
BAR
THE DON
CUT ME
BEERS
Happy Hour
7pm - 8pm

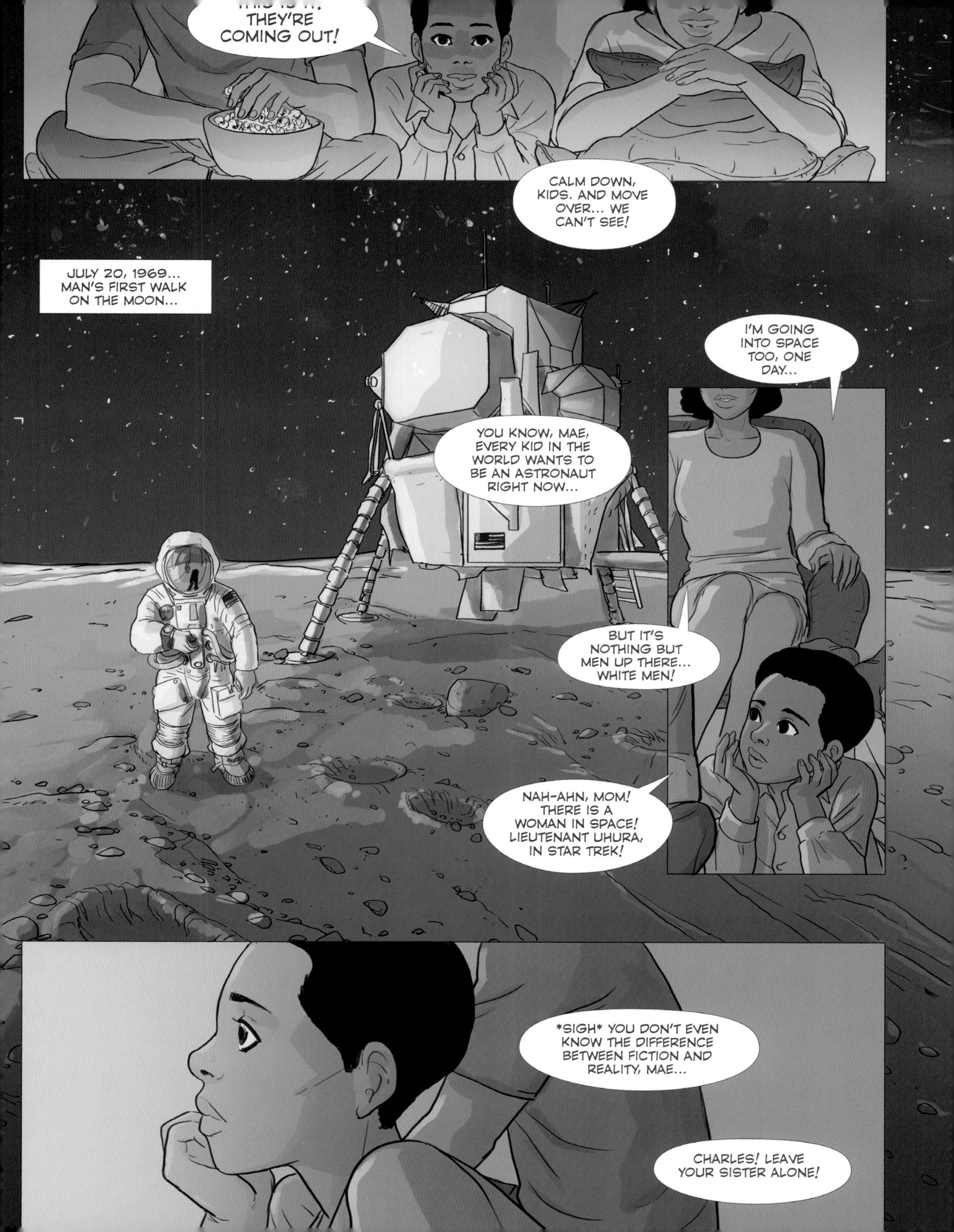
THEY'RE COMING OUT!
CALM DOWN, KIDS. AND MOVE OVER... WE CAN'T SEE!
JULY 20, 1969... MAN'S FIRST WALK ON THE MOON...
I'M GOING INTO SPACE TOO, ONE DAY...
YOU KNOW, MAE, EVERY KID IN THE WORLD WANTS TO BE AN ASTRONAUT RIGHT NOW...
BUT IT'S NOTHING BUT MEN UP THERE... WHITE MEN!
NAH-AHN, MOM! THERE IS A WOMAN IN SPACE! LIEUTENANT UHURA, IN STAR TREK!
SIGH YOU DON'T EVEN KNOW THE DIFFERENCE BETWEEN FICTION AND REALITY, MAE...
CHARLES! LEAVE YOUR SISTER ALONE!

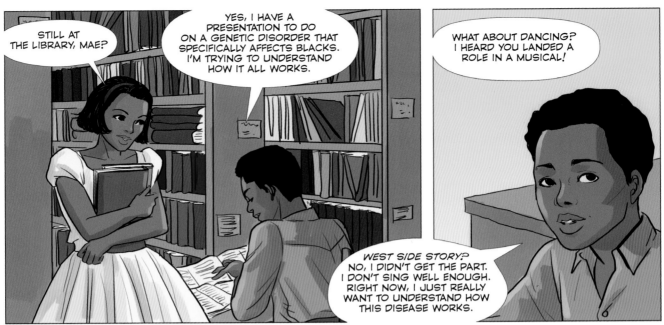

STILL AT THE LIBRARY, MAE?

YES, I HAVE A PRESENTATION TO DO ON A GENETIC DISORDER THAT SPECIFICALLY AFFECTS BLACKS. I'M TRYING TO UNDERSTAND HOW IT ALL WORKS.

WHAT ABOUT DANCING? I HEARD YOU LANDED A ROLE IN A MUSICAL!

WEST SIDE STORY? NO, I DIDN'T GET THE PART. I DON'T SING WELL ENOUGH. RIGHT NOW, I JUST REALLY WANT TO UNDERSTAND HOW THIS DISEASE WORKS.

MAE JEMISON, ON BEHALF OF THE CITY OF CHICAGO, I'M THRILLED TO AWARD YOU FIRST PRIZE IN THE SCIENCE COMPETITION FOR YOUR PRESENTATION. CONGRATULATIONS!

I HAVE GOOD NEWS: THE COMMITTEE HAS DECIDED TO FUND A NATIONAL MERIT GRANT IN YOUR NAME.

THIS IS BEYOND OUR WILDEST DREAMS, DOROTHY. A SCHOLARSHIP TO STANFORD! ONE OF THE TOP UNIVERSITIES IN THE COUNTRY!

BUT SHE'S ONLY SIXTEEN, CHARLIE! AND IT'S ALL THE WAY ON THE WEST COAST...

SIXTEEN AND ALL ALONE... IN CALIFORNIA? CAN SHE DO IT?

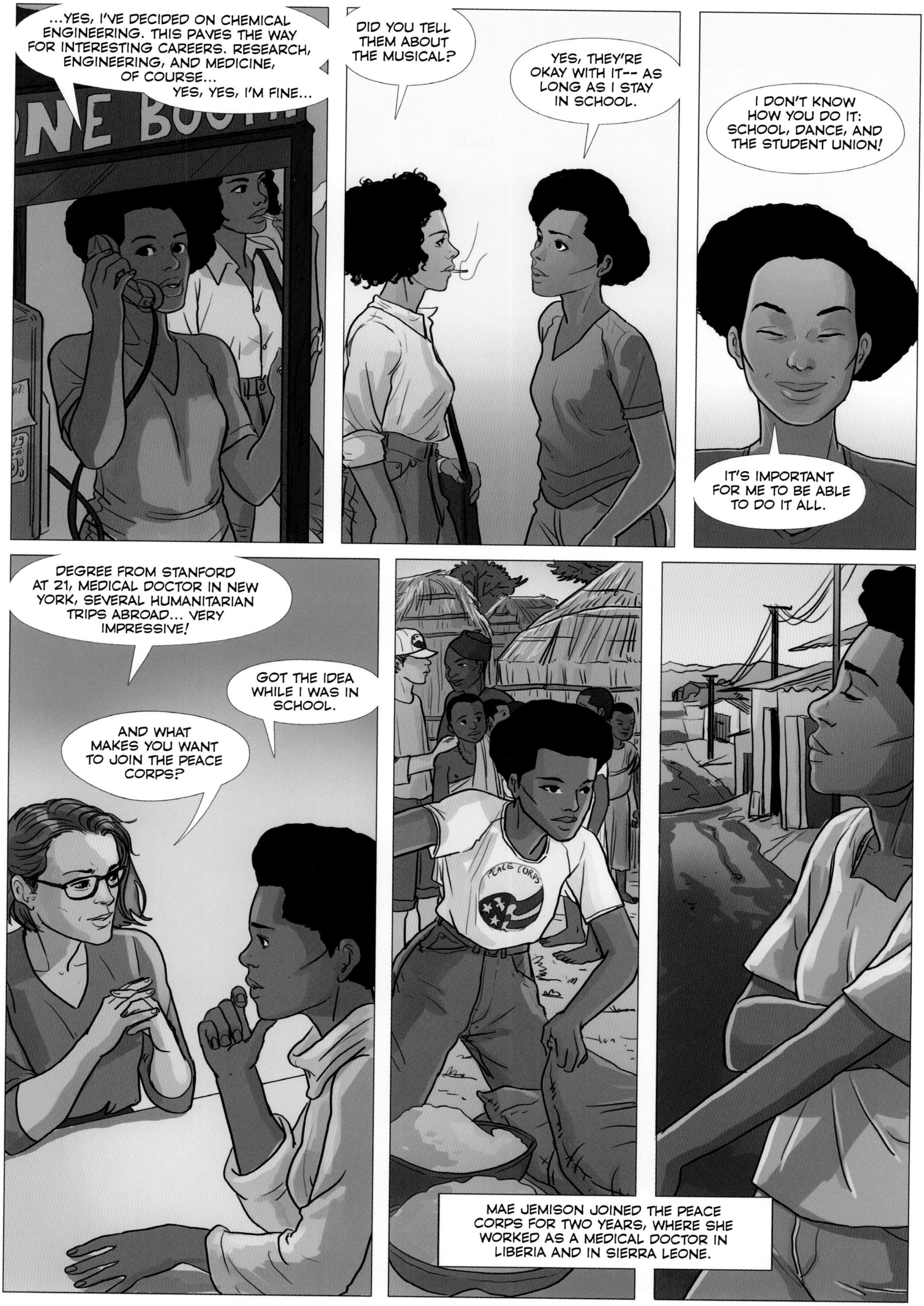

...YES, I'VE DECIDED ON CHEMICAL ENGINEERING. THIS PAVES THE WAY FOR INTERESTING CAREERS. RESEARCH, ENGINEERING, AND MEDICINE, OF COURSE... YES, YES, I'M FINE...
ONE BOOTH
DID YOU TELL THEM ABOUT THE MUSICAL?
YES, THEY'RE OKAY WITH IT-- AS LONG AS I STAY IN SCHOOL.
I DON'T KNOW HOW YOU DO IT: SCHOOL, DANCE, AND THE STUDENT UNION!
IT'S IMPORTANT FOR ME TO BE ABLE TO DO IT ALL.
DEGREE FROM STANFORD AT 21, MEDICAL DOCTOR IN NEW YORK, SEVERAL HUMANITARIAN TRIPS ABROAD... VERY IMPRESSIVE!
AND WHAT MAKES YOU WANT TO JOIN THE PEACE CORPS?
GOT THE IDEA WHILE I WAS IN SCHOOL.
PEACE CORPS
MAE JEMISON JOINED THE PEACE CORPS FOR TWO YEARS, WHERE SHE WORKED AS A MEDICAL DOCTOR IN LIBERIA AND IN SIERRA LEONE.

IN 1983, SALLY RIDE BECAME THE FIRST FEMALE ASTRONAUT TO GO INTO SPACE.

Sally's riding high

TO FOLLO

...COME TO THINK OF IT, WHY DON'T I APPLY TOO?

AS A CHILD, I KNEW ALL THE APOLLO PROGRAMS BY HEART. I GRADUATED FROM STANFORD AT 21, I'M A DOCTOR...

YOU CAN DO ANYTHING, GIRL! NEVER FORGET THAT!

BECOMING AN ASTRONAUT ISN'T EASY. IT'S A LONG ROAD FILLED WITH SETBACKS...

PATIENCE AND TENACITY ARE REQUIRED QUALITIES.

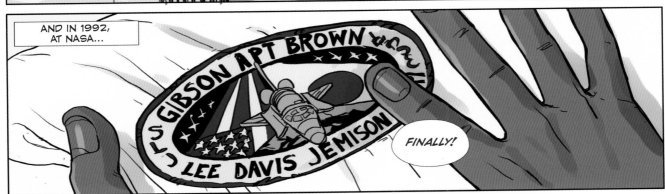

AND IN 1992, AT NASA...

GIBSON APT BROWN LEE DAVIS JEMISON

FINALLY!

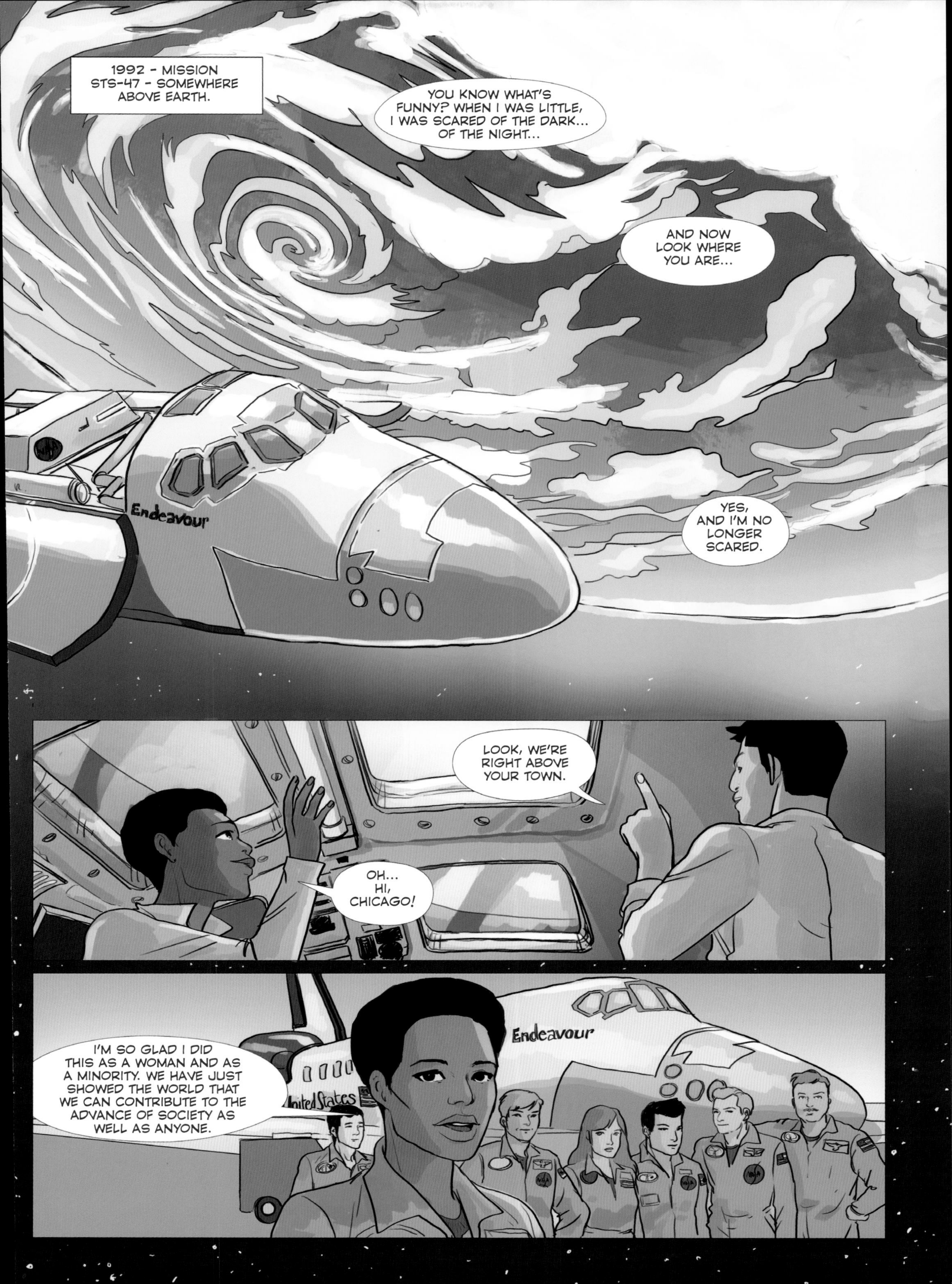
1992 - MISSION STS-47 - SOMEWHERE ABOVE EARTH.
YOU KNOW WHAT'S FUNNY? WHEN I WAS LITTLE, I WAS SCARED OF THE DARK... OF THE NIGHT...
AND NOW LOOK WHERE YOU ARE...
YES, AND I'M NO LONGER SCARED.
Endeavour
LOOK, WE'RE RIGHT ABOVE YOUR TOWN.
OH... HI, CHICAGO!
I'M SO GLAD I DID THIS AS A WOMAN AND AS A MINORITY. WE HAVE JUST SHOWED THE WORLD THAT WE CAN CONTRIBUTE TO THE ADVANCE OF SOCIETY AS WELL AS ANYONE.
Endeavour
United States

A FEW YEARS LATER...

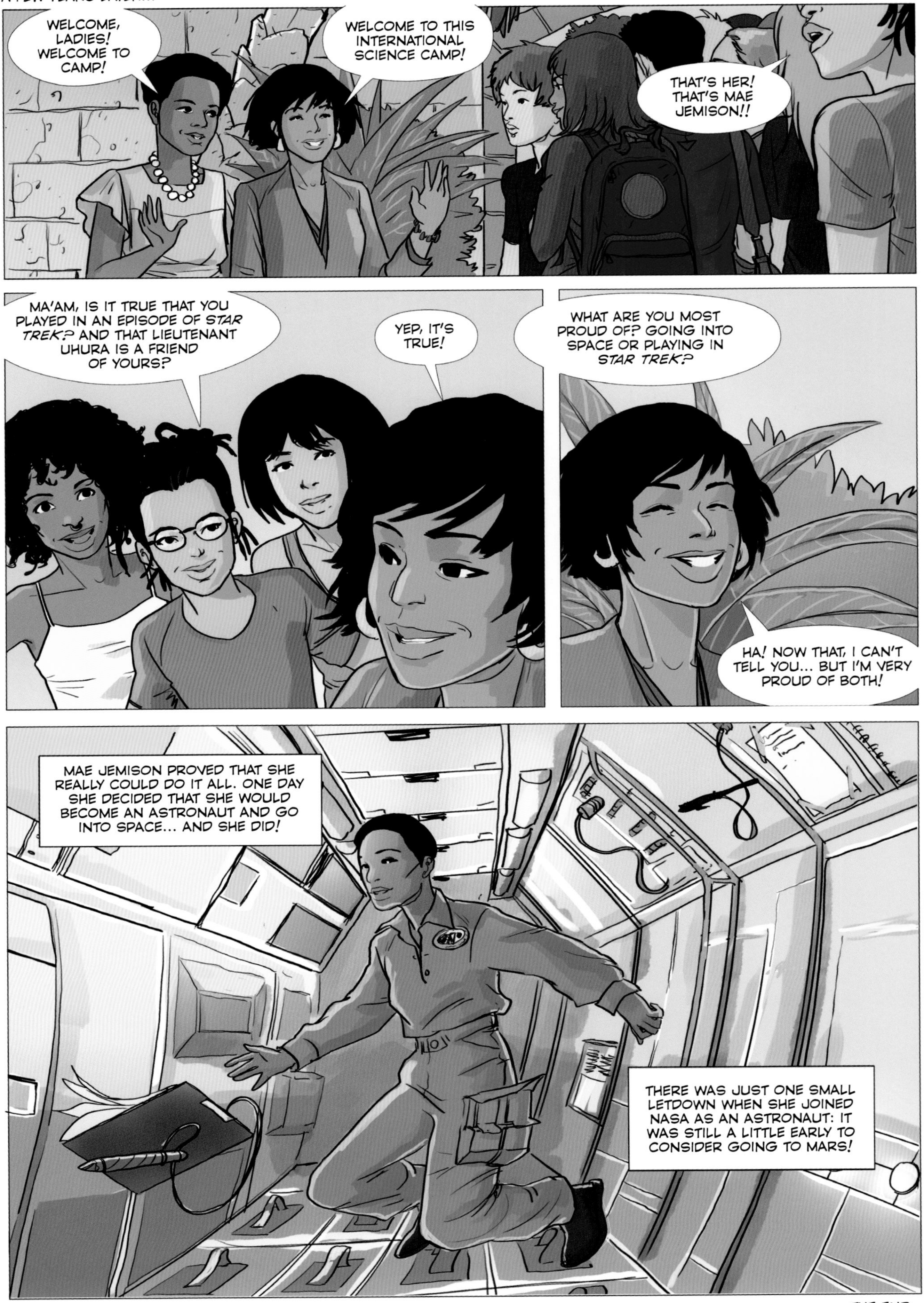
WELCOME, LADIES! WELCOME TO CAMP!
WELCOME TO THIS INTERNATIONAL SCIENCE CAMP!
THAT'S HER! THAT'S MAE JEMISON!!
MA'AM, IS IT TRUE THAT YOU PLAYED IN AN EPISODE OF *STAR TREK*? AND THAT LIEUTENANT UHURA IS A FRIEND OF YOURS?
YEP, IT'S TRUE!
WHAT ARE YOU MOST PROUD OF? GOING INTO SPACE OR PLAYING IN *STAR TREK*?
HA! NOW THAT, I CAN'T TELL YOU... BUT I'M VERY PROUD OF BOTH!
MAE JEMISON PROVED THAT SHE REALLY COULD DO IT ALL. ONE DAY SHE DECIDED THAT SHE WOULD BECOME AN ASTRONAUT AND GO INTO SPACE... AND SHE DID!
THERE WAS JUST ONE SMALL LETDOWN WHEN SHE JOINED NASA AS AN ASTRONAUT: IT WAS STILL A LITTLE EARLY TO CONSIDER GOING TO MARS!

THE END

Stephanie Kwolek
1923 – 2014
American chemist

Stephanie Kwolek was born in Pennsylvania in 1923 to a family of Polish extraction. Her father, a naturalist, died when she was ten years old, but not before sharing his love of trees and nature with his daughter. Her mother, a seamstress, gave her a taste for fashion, a field that Stephanie could have chosen as a career.

But she eventually opted for science instead, and graduated from Carnegie Mellon University's Margaret Morrison Carnegie College, a women's college, with a degree in chemistry.

To finance her studies, she landed a small job at Dupont, an innovative chemistry company that revolutionized stockings with the invention of nylon in the 1930s.

It was while working at that company that she looked for a way to make car tires stronger and discovered Kevlar, a particularly strong fiber.

Five times stronger than steel and extremely lightweight, it is used today for bulletproof vests, airplanes, ships, appliances, and for just about anything that requires extreme resistance.

The multiple award-winning Stephanie Kwolek was inducted into the National Women's Hall of Fame in 2003.

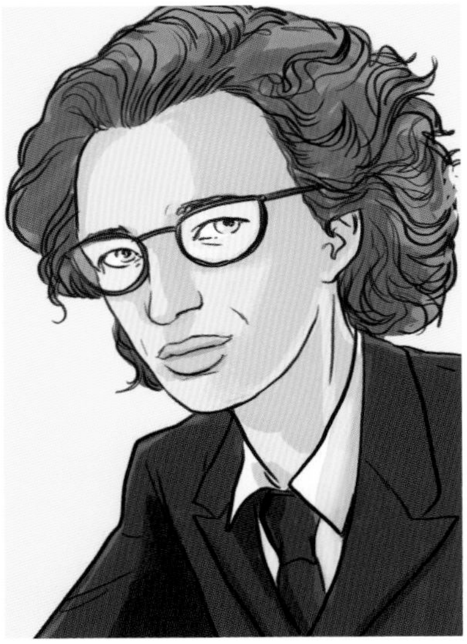

Grace Murray Hopper
1906 - 1992

American computer scientist
Presidential Medal of Freedom

Grace Murray Hopper was born in 1906 in New York to a military family.

Passionate about mathematics, she earned a PhD in mathematics from Yale University and then taught the subject at Vassar College, where she was an associate professor for several years before she enlisted in the Navy, inspired by the family military tradition. Up until she retired from the U.S. Navy at the age of 60, she was the oldest officer there and held the high rank of Rear Admiral.

As someone who enjoyed thinking outside the box, she kept the clock in her office running counter clockwise in order to remind herself that she was there to change the way people thought.

This mindset is what led her to question why computers could only do arithmetic and only expressed themselves with symbols. Imagining that everyone would one day have access to this technology, she sought to go beyond that function and in 1959 came up with a much more accessible computer programming language: COBOL.

In 2016, Barack Obama posthumously awarded her the Presidential Medal of Freedom.

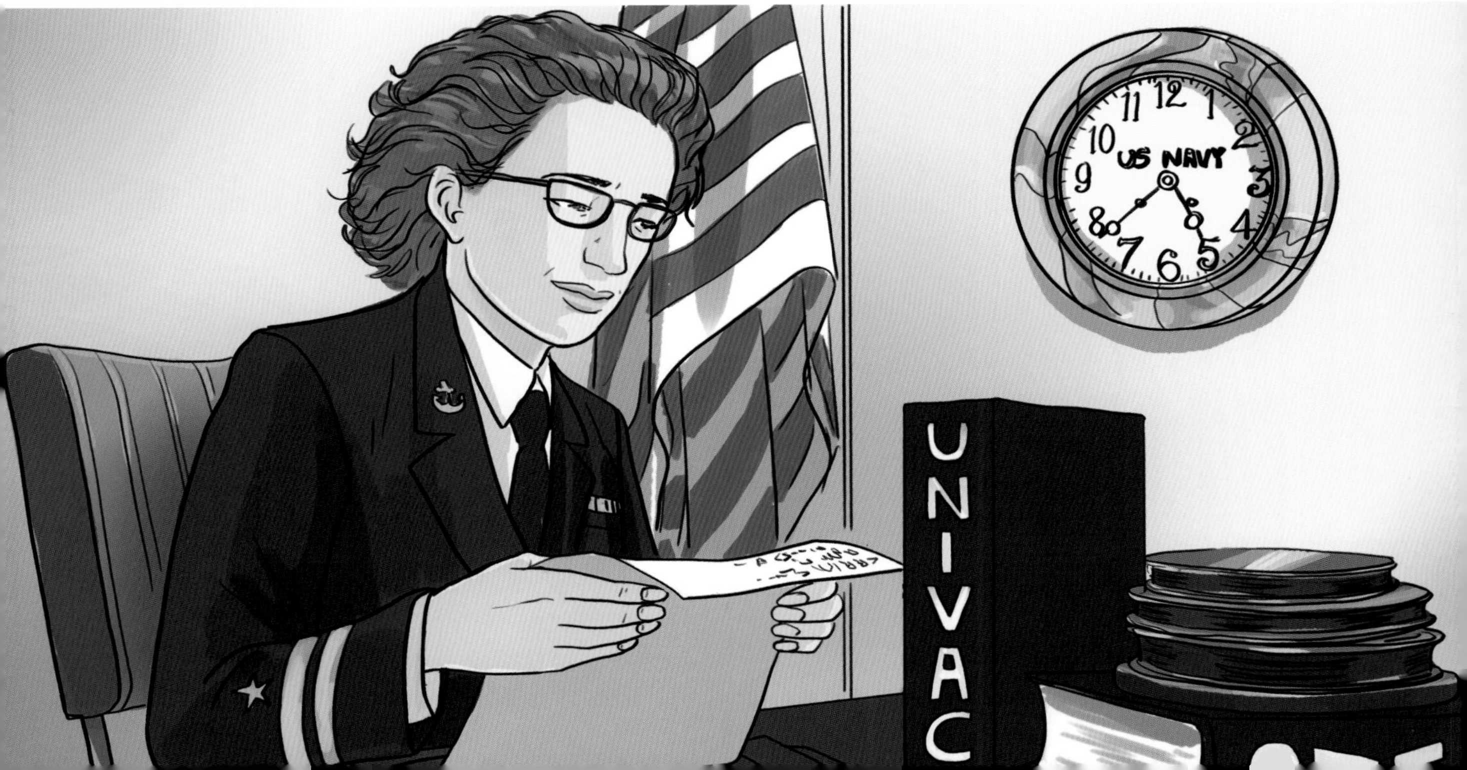
US NAVY
UNIVAC

Xie Yi
1967 –

Chinese chemist
2015 L'Oréal-Unesco Women and Science Award

Born in 1967 in Fuyang, in the Anhui province in Eastern China, **Xie Yi** earned a degree in chemistry in 1988 from Xiamen University, in the province of Fujian.

She started her career as an assistant engineer in a chemical factory before resuming her studies, this time at the University of Science and Technology of China, where she earned a PhD. Her research focused on nanotechnology, and in particular on thermoelectricity, the harnessing and transformation of solar energy into electricity. She was also dedicated to the fight against pollution.

In 2013, Yi was elected fellow of the Royal Society of Chemistry (United Kingdom) and fellow of the Chinese Academy of Sciences

In 2014, she received the TWAS (Third World Academy of Sciences) scientific award, and in 2015, she received a L'Oréal-Unesco Women and Science Award.

Also available from NBM Comics Biographies:
The Beatles in Comics
The Rolling Stones in Comics
Bob Marley in Comics
The Disney Bros.
Elvis
Einstein
Michael Jackson in Comics

See previews, get exclusives and order from:
NBMPUB.COM
We have hundreds of graphic novels
available including many more bios.
Subscribe to our monthly newsletter
Follow us on Facebook & Instagram
(nbmgraphicnovels), Twitter (@nbmpub).

NBM
160 Broadway, Suite 700, East Wing
New York, NY 10038
Catalog upon request

THE BEATLES
in comics
nbm

The ROLLING STONES
in comics
nbm

BOB MARLEY
in comics
nbm

THE DISNEY BROS.
The Fabulous Story of Walt and Roy
nbm

Elvis
nbm

A. Einstein
THE POETRY OF REAL
E=MC²
nbm

MICHAEL JACKSON
in comics
nbm

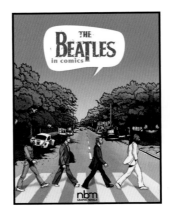

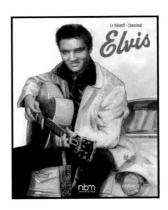

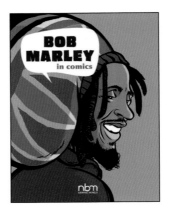

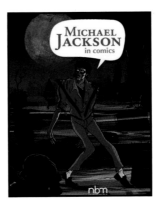

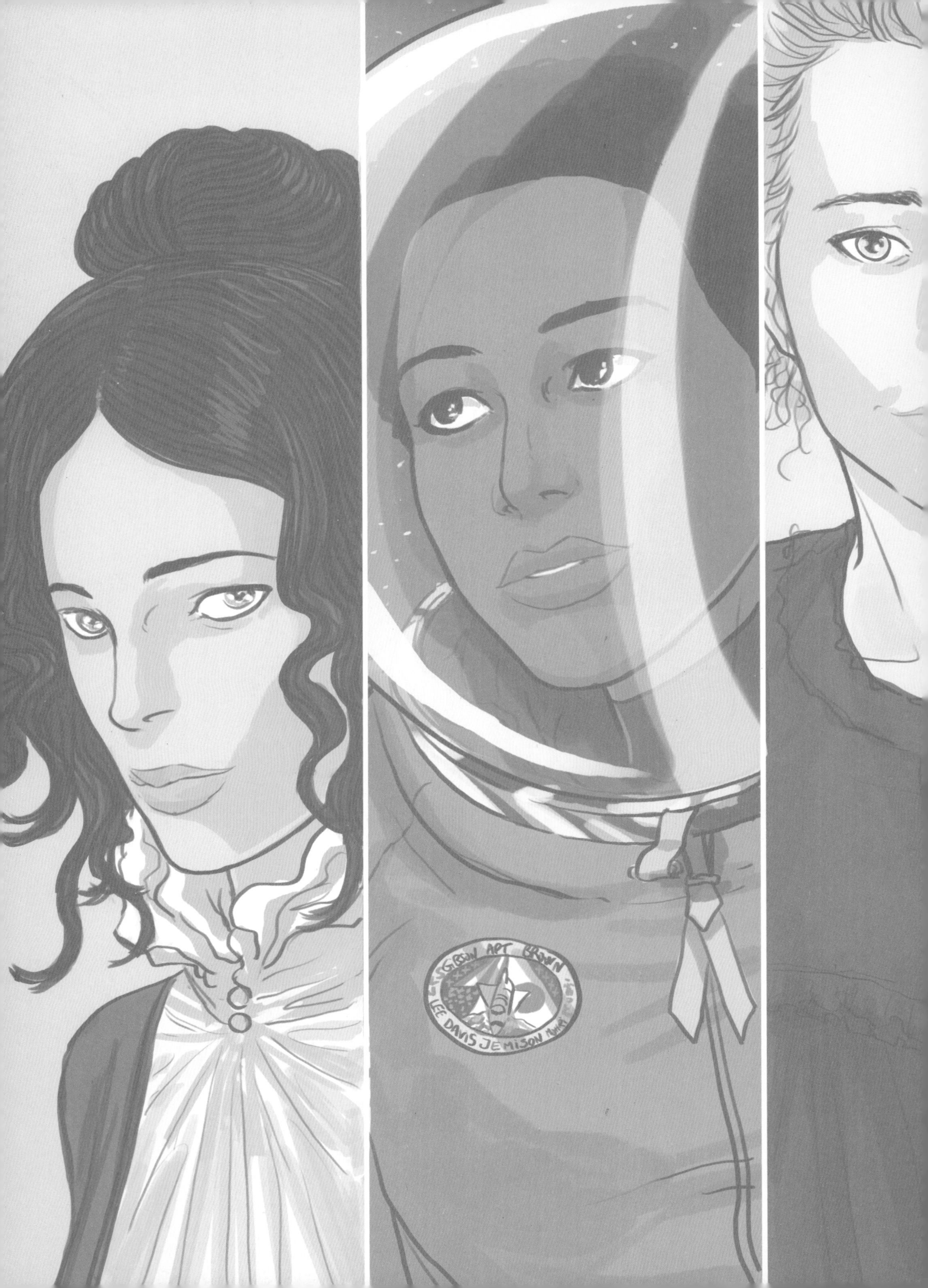
GIBSON APT BROWN
LEE DAVIS JEMISON